Chemistry Connections
Second Edition

The Complementary Science Series is an introductory, interdisciplinary, and relatively inexpensive series of paperbacks for science enthusiasts. These titles cover topics that are particularly appropriate for self-study although they are often used as complementary texts to supplement standard discussions in textbooks. They are deliberately unburdened by excessive pedagogy, which is distracting to many readers, and avoid the often plodding treatment in introductory texts. The series was conceived to fill the gaps in the literature between conventional textbooks and monographs by providing real science at an accessible level, with minimal prerequisites so that students at all stages can have expert insight into important and foundational aspects of current scientific thinking.

Many of these titles have strong interdisciplinary appeal, such as a chemist writing about applications of biology to physics, or vice versa, and all have a place on the bookshelves of literate laypersons. Potential authors are invited to contact our editorial office at www.academicpressbooks.com.

Complementary Science Series

ACADEMIC PRESS
An imprint of Elsevier Science

Earth Magnetism
Wallace Hall Campbell

Physics in Biology and Medicine, 2nd Edition
Paul Davidovits

Fundamentals of Quantum Chemistry, 2nd Edition
J.E. House

Mathematics for Physical Chemistry, 2nd Edition
Robert Mortimer

The Physical Basis of Chemistry, 2nd Edition
Warren S. Warren

Introduction to Relativity
John B. Kogut

www.academicpressbooks.com

Senior Publishing Editor, Physical Sciences	Jeremy Hayhurst
Senior Project Manager	Angela Dooley
Editorial Coordinator	Nora Donaghy
Marketing Manager	Linda Beattie
Cover Design	G.B.D. Smith
Copyeditor	Charles Lauder, Jr.
Composition	International Typesetting Co.
Printer	Maple-Vail

This book is printed on acid-free paper. ∞

Series logo: Images © 2000 PhotoDisc, Inc.
Front cover image: Aurora borealis. Photo Courtesy of George Lepp/Tony Stone Images
More detail on auroral displays can be found in Campbell, *Earth Magnetism*, Complimentary Science Series, 2001

Academic Press
An imprint of Elsevier Science
525 B Street, Suite 1900, San Diego, California 92101-4495, USA
http://www.academicpress.com

Academic Press
84 Theobald's Road, London WC1X 8RR, UK
http://www.academicpress.com

Academic Press
200 Wheeler Road, Burlington, Massachusetts 01803, USA
www.academicpressbooks.com

Library of Congress Catalog Card Number: 2003102993

International Standard Book Number: 0-12-400151-3

PRINTED IN THE UNITED STATES OF AMERICA
03   04   05   06   9   8   7   6   5   4   3   2   1

Chemistry Connections
Second Edition

The Chemical Basis of Everyday Phenomena

Kerry K. Karukstis
Gerald R. Van Hecke

Department of Chemistry
Harvey Mudd College
Claremont, California

ACADEMIC PRESS

An imprint of Elsevier Science

Amsterdam Boston London New York Oxford Paris
San Diego San Francisco Singapore Sydney Tokyo

The important thing is not to stop questioning.
—Albert Einstein

To our parents

Contents

Preface   **xiii**

Conventions   **xvii**

1 Connections to Atomic Properties   **1**
Why Do Iodine and Barium Enhance CAT Scans?   1
Why Does a Kitchen Gas Burner Glow Yellow When a Pot of Boiling
    Water Overflows?   2

2 Connections to Gases   **5**
What Causes an Egg to Crack if It's Boiled Too Rapidly?   5
When a Flambé Is Prepared, Why Is the Liqueur Heated Prior to Light-
    ing the Flame?   7
How Do Correction Fluids like Liquid Paper and White-Out Work? . . .   8
Why Do Cosmetic Cold Creams Feel Cool When Applied to the Skin? .   9

3 Connections to Solutions   **11**
Why Do Carbonated Drinks Go Flat as They Warm?   11
Why Do Crystals Form in Wine over Time?   13
What Makes a Lava Lamp Work?   14
How Is the Toxicity of Barium Sulfate Controlled for X-Rays?   17
Why Is Milk of Magnesia an Antacid?   18
Why Are Ice Cubes Cloudy on the Inside?   20
Why Are Metal or Glass Bowls Best for Whipping Egg Whites?   21
How Do Fog Machines Create the Artificial Fog or Smoke Used in
    Theatrical Productions?   23
Why Are Opals and Pearls Iridescent?   25

4 Connections to Solids   **29**
Why Does Morton Salt Claim "*When It Rains It Pours*"?   29
What Are "Desiccants" and Why Are They in Packaged Products? . . .   30
Why Is the Hope Diamond Blue?   32

5 Connections to Chemical Reaction Types   **35**

Why Is There Abundant White Smoke from the Space Shuttle Booster
    Rockets on Lift-Off? 35
Why Do Homemade Copper Cleaners Use Vinegar? 36
Why Is "Fighting" Magnesium Fires with Water or CO_2 Dangerous? . . 39
Why Is Hydrogen Peroxide Kept in Dark Plastic Bottles? 40
Why Does a Flashbulb Develop a White Coating after a Flash? 41
Why Did Dorothy Have to Oil the Tin Woodman in *The Wizard of Oz*? . 43
What Puts the "Blue" in "Blue Jeans"? 44
How Is Lime Used to Mitigate the Acid Rain Problem? 46
Why Are Bombardier Beetles Known as "Fire-Breathing Dragons"? . . 48
Why Do Seashells Vary in Color? 50
Why Is Vinegar Recommended for Cleaning Automatic Coffee Makers
    and Steam Irons? 52
Why Does Soap Scum Form? Why Are Phosphates in Detergents? . . . 53
Why Do Old Paintings Discolor? 56
How Do Hair Coloring Products "Remove" Gray in Hair? 58
What Is the Source of Dr. Seuss' Green Eggs (and Ham)? 59
How Does Washing Soda Soften Hard Water? 60
Why Does Baking Soda Extinguish a Fire? 62
What Is the Origin of the Expression "in the Limelight"? 64
What Causes Puff Pastry to Expand? 66
Why Does Chlorine in Swimming Pools Work Best at Night? 69

6 Connections to Acids and Bases   **72**

What Does pH Stand For? 72
Why Does Disappearing Ink Disappear? 73
How Are Light Bulbs Frosted? 75

7 Connections to Intermolecular Forces   **78**

Why Are General Anesthetics Administered as Gases? 78
Why Won't Water Relieve the Burning Sensation of Chili Peppers? . . . 83
Why Are Bubbles in a Carbonated Drink Spherical? 84
How Do Furniture Polishes Repel Dust? 85
How Do Windshield Coatings Improve Visibility in a Rainstorm? 87
How Is a Fabric Made Water-Repellent or Waterproof? 89
How Does a Bullet-Proof Vest Work? How Is It Made? 92
Why Is Cotton so Absorbent and Why Does It Dry So Slowly? 95
What Makes a No-Tears Shampoo? 97
How Does Milk Froth for Cappuccino Coffee? 100
Why Is Sickle Cell Anemia a "Molecular Disease"? 102

8 Connections to Phases and Phase Transitions   **105**

What Do Meteorologists Use to "Seed" Clouds? 105

Why Do Citrus Growers Spray Their Trees with Water to Protect Them
from a Freeze? 107
What Is the Dark Spot Left on the Inside of a Light Bulb When it Burns
Out? 109
What are the Red or Silver Liquids in Thermometers? 111
What Nineteenth-Century "Disease" Destroyed Cathedral Organ Pipes? 113

9 Connections to Equilibria   116
What Causes the Fizz When an Antacid Is Added to Water? 116
What Is an Antidote for Cyanide Poisoning? 118
Why Is EDTA Added to Salad Dressings? 120
Why Do Hydrangeas Vary in Color When Grown in Dry versus Wet
Regions? 122
Why Are Floor Waxes Removable with Ammonia Cleansers? 124
How Do Drano and Liquid-Plumr Unclog Drains? 126

10 Connections to Thermodynamics   128
Why Do CAT Scans Often Cause a "Warm Flush" Sensation? 128
What Provides the Insulating Qualities in Double- and Triple-Pane Win-
dows? 130
How Does a Snow-Making Machine Work? 131
Why Does a Plaster Cast Warm as the Cast Hardens? 133
What Causes an Instant Ice-Pack to Cool? 134

11 Connections to Kinetics   137
Why Do Some Batteries Last Longer When Stored in a Refrigerator? . . 137

12 Connections to Light   139
Why Do Lightsticks Glow? 139
Why Is an Astronaut's Visor So Reflective? 143
Why Is the Aurora Borealis So Colorful? 145
What Causes the Pearlescent Appearance of Some Paints? 147
Why Is It Incorrect to Call U.S. Bills "Paper Currency"? 149
What Is the Purpose of the Thread That Runs Vertically through the
Clear Field on the Face Side of U.S. Currency? 152
Why Do U.S. Bills Shift in Color When Viewed from Different Angles? 155
What Is an Optical Brightener? 158
What Is the Difference between a Sunscreen and a Sunblock? 161

13 Connections to Organic Chemistry   166
How Do Sutures Dissolve? 166
How Do You Deodorize Skunk Spray on Your Dog? 168
How Do Forensic Chemists Use Visible Stains to Trap Thieves? 170
Why Is White Willow Bark Known as "Nature's Aspirin"? 172
Why Does Balsamic Vinegar Have a Sweet Taste? 175

How Do "Sniffing Dogs" Detect Narcotics or Explosives? 176
Why Should You Avoid Acidic Foods When Taking Penicillin? 178
Why Do We Detect Odor in an "Odorless Gas" Leak? 180
Why Must Ammonia Cleansers Never Be Mixed with Bleach? 181
What Is "Alpha Hydroxy Acid" in Antiaging Creams? 183
What Is an "Alcohol-Free" Cosmetic? 186
What Is the "Cool Sensation" in Toothpaste and Breath Fresheners? . . 189
When We Buy Fresh Fish, Why Does Its Smell Indicate Its Freshness? . 193
Are Flamingos Naturally Pink? 195
What Causes Ice Cream to Develop a Gritty Texture during Long Peri-
    ods of Storage? 197
Why Is Some Olive Oil Designated as "Extra Virgin"? 199
How Can Sucralose, an Artificial Sweetener Made from Sugar, Contain
    No Calories? 205

14 Connections to Polymers   208
How Does a Timed-Release Medicine Work? 208
How Does "Scratch & Sniff" and Carbonless Copy Paper Work? 211
How Do "Repositional Self-Adhesive Paper Notepad Sheets" (Post-it
    Notes) Work? 214
What Is "Shatterproof" Glass? 216
Why Does Superglue Stick to Almost Every Surface? 219
What Is the Difference between Hard and Soft Contact Lenses? 221

15 Connections to Materials   226
What Is the Composition of an Artificial Hip? 226
What Is Liquidmetal (in Liquidmetal Golf Clubs)? 228
How Does an Automatic Fire Sprinkler Head Operate? 230

16 Connections to Biochemistry   233

17 Connections to the Environment   234

Index   235

Preface(s)

SECOND EDITION

The very favorable response to *Chemistry Connections* encouraged us to prepare this expanded second edition. We have improved the collection of questions, particularly to enhance the number of applications in the area of organic chemistry. Furthermore, we have revised the organization of questions to assist instructors in selecting examples to coordinate with the principles and topics covered in their chemistry classes.

The new organization, photographs, and chapter headings will stream-line the connection with a range of general chemistry textbooks and courses. For the general reader, however, our intention is still to demonstrate the wide scope and significance of chemistry and the ever-present connection of the discipline to our daily lives.

Kerry K. Karukstis
Gerald R. Van Hecke

Conventions

The answers to many questions include chemical formulas shown in common drawing notation. For readers unfamiliar with such notation the examples below should help to interpret the figures. A line represents a chemical bond. The number of lines between two atoms represents the number of chemical bonds between the joined atoms. Unless stated otherwise, all lines join two carbon atoms. The junction of two lines implies the presence of a carbon atom, unless another atom is shown. Each carbon should have four lines drawn to it. Any missing line should be interpreted as a bond to a hydrogen atom. For example, a line that terminates should be viewed as a carbon bonded to three hydrogens and bonded to a fourth atom shown or implied. In this edition care has been taken to better illustrate the true three-dimensional structures that molecules represent. To this end, a solid wedge means the bond is coming out of the plane of the page, and a dashed wedge implies the bond points below the printed page. See the drawings below for examples.

FIRST EDITION

Chemistry Connections: The Chemical Basis of Everyday Phenomena highlights the fundamental role of chemical principles in governing our everyday experiences and observations. This collection of contemporary real-world examples of chemistry in action is written in a question-and-answer format with presentations in both lay and technical terms of the chemical principles underlying numerous familiar phenomena and topical curiosities. Introductory college chemistry students and educators as well as laypersons with an inquisitiveness about the world around them will find the book an informative introduction to the context of chemistry in their lives.

Assessment of the Need for this Book

United States Education Secretary Richard W. Riley recently commented on the results of a national assessment of scientific literacy among U.S. high school graduates.[1] "We are confronted by a paradox of the first order. We Americans are fascinated by technology. Yet, at the same time, Americans remain profoundly ignorant."

Former National Science Foundation director Neal Lane concurs: "I have become especially conscious of the discrepancy between the public's interest in, even fascination with, science and its limited knowledge about scientific concepts and issues."[2] He further adds, "All scholarly fields—poetry or philosophy, architecture or agriculture—suffer from their separation from the public, although in the case of science, the separation may be more extreme. And yet science and the technology it spawns pervade the very structure of everyone's life...".

Perhaps no scientific field is less understood and appreciated by the public, and in particular by students, than chemistry. A general misunderstanding of the nature of chemistry and even the meaning of the word chemical pervades our society.

Students and the general public alike are further unaware of the broad scope of chemistry and the impact of the discipline on many fields. Conveying the importance and relevance of chemistry to our world is one of the greatest challenges facing chemists and chemical educators today.

The revitalization of chemistry education has received much recent attention and taken many forms. Modes of teaching, textbooks, laboratory instruction—all aspects of the chemistry curriculum have undergone scrutiny for reform. A recent National Science Foundation report, *Shaping the Future: New Expectations for Undergraduate Education in Science, Mathematics, Engineering, and Technology*[3] characterizes the nature of the most successful curricular and pedagogical improvements:

"A simple precis is that these improvements are attempting to nurture a sense of wonder among students about the natural world, to maintain students' active curiosity about this world while equipping them with tools to explore it and to learn." Indeed, a recent survey of college chemistry courses revealed findings that indicated an increased emphasis towards the presentation of chemical principles reflecting "the more 'relevant' chemistry of 'everyday living'."[4] These initiatives are based on the development of a curriculum that engages a broad base of students and that provides students with a familiar context for chemical concepts, stimulating their desire to explore further.

Approach Used in this Text

By what mechanism do chemistry textbooks and monographs demonstrate the *relevance* of chemistry? Most general chemistry texts include short features highlighting "real world" examples of the various chemical principles that are illustrated within each chapter. Some texts take a more revolutionary approach to promote the interest of students in chemistry by a text book structure that focuses on key household products (e.g., food, apparel) and technologies or industries (e.g., health, communications, transportation) as a means of introducing the chemical principles of a standard college curriculum. Other tradebooks are broad overviews of the global impact of chemistry on society, usually written in non-technical language.

In *Chemistry Connections* we have adopted a separate approach, collecting in one volume an assortment of provocative, topical questions that are raised by our everyday experiences and that are answered by the application of chemical principles.

The design of the book makes it compatible with any general chemistry text for students and educators or suitable as an independent book for any individual with a curiosity about the world around them. From the reader's viewpoint, the pertinence of chemistry to each question ranges from straightforward examples to more intriguing applications. We chose the question-and-answer format to provide a motivating force to interest the reader to learn the necessary chemical principles to understand everyday phenomena. Explanations are provided in both lay and technical terms—an initial description to satisfy the curious reader, followed by a more in-depth account to underscore the chemical nature of the phenomenon. We expect that readers will also quickly appreciate that an interplay of several chemical principles is often needed to explain fully a real-world observation, a realization too often overlooked by the beginning student or casual readers. Each question is indexed according to key principles or terms to provide teachers with the flexibility to select pertinent examples for class discussion. To furnish readers with additional related information for further exploration, we chose to focus on references to Web sites. With today's increasing access to the Internet, these selections may be more readily available than many hardcopy references. We recognize the transient nature of the World Wide Web, however, and encourage the reader to use these sites as starting points for their own discovery of related electronic materials.

Kerry K. Karukstis
Gerald R. Van Hecke

References

[1] E. Woo, *Los Angeles Times,* May 3, 1997.

[2] *The Chronicle of Higher Education,* December 6, 1996, p. A84.

[3] Shaping the Future: New Expectations for Undergraduate Education in Science, Mathematics, Engineering, and Technology, Directorate for Education and Human Resources, 1996.

[4] H. L.Taft, *Journal of Chemical Education* 74, 595–599 (1997).

Chapter 1

Connections
to Atomic Properties

| 1.1 | ## Why Do Iodine and Barium Enhance CAT Scans? |

The innovative and technological achievement of computerized axial tomography required a sophisticated understanding of physiology, mathematics, and radiology. A clear comprehension of chemical principles also contributes to the success of this diagnostic technique.

The Chemical Basics

The technique of computerized axial tomography (CAT) assists medical examiners in viewing the internal organs of the body. The technique has been widely used since its development in the 1970s by the British electrical engineer Sir Godfrey Hounsfield and the South African-born U.S. physicist Allen Cormack. These scientists won the Nobel Prize for Physiology or Medicine in 1979 for their contributions to the development of this diagnostic technique.[1]

In a CAT scan, cross-sectional images are generated using X-rays directed through the body using a rotating tube. X-rays not absorbed by the body reach a radiation detector where the signals are integrated to produce an image that assesses the density of tissues at various locations. X-rays are absorbed differentially — with denser objects such as bones absorbing extensively and soft tissues such as blood vessels absorbing relatively few X-rays. Thus bones appear as light areas on the image, soft tissues as dark regions.

What can one do to image specific soft tissues using X-rays? One must enhance the density of these regions to decrease the ability to transmit X-rays through the region. To aid in the creation of such images or CAT scans, *contrast media* are often ingested or injected into the body. Contrast medium fluids that are opaque (i.e., not transparent) to X-rays are known as *radiopaque media*. These media highlight the areas of the body being scanned as a consequence of the inability of X-rays to penetrate these substances. Media containing barium or iodine meet these criteria. For imaging gastrointestinal tracts, barium-based media are orally ingested; for visualizing major blood vessels, iodinated contrast media are injected into the patient's veins using intravenous access.

The Chemical Details

Why are barium- and iodine-based materials selected for contrast media? The production of X-ray images depends on the differences between the X-ray absorbing power of various tissues. This difference in absorbing power is called *contrast* and is directly dependent on tissue density. To artificially enhance the ability of a soft tissue to absorb X-rays, the density of that tissue must be increased. The absorption by targeted soft tissue of aqueous solutions of barium sulfate and iodized organic compounds provides this added density through the heavy metal barium and the heavy nonmetal iodine.

KEY TERMS: X-ray   contrast

References

[1] "Press release: The 1979 Nobel Prize in Physiology or Medicine." Nobelforsamlingen, Karolinska Institutet. The Nobel Assembly at the Karolinska Institute. 11 October 1979, http://www.nobel.se/medicine/laureates/1979/press.html

Related Web Sites

▶ "Computed Tomography." Radiological Society of North America, Inc. (RSNA) Center, Chicago, http://www.radiologyinfo.org/content/ct_of_the_body.htm

1.2   Why Does a Kitchen Gas Burner Glow Yellow When a Pot of Boiling Water Overflows?

A cook recognizes the tell-tale signs of an overflowing pot of boiling water — the characteristic hiss as the water hits the hot gas burner and evaporates and the familiar accompanying yellow glow. Chemistry on the atomic level is responsible for the brilliant yellow illumination.

The Chemical Basics

The yellow color imparted to a natural gas flame originates from the ignition of sodium atoms or ions. The common source of sodium is salt (sodium chloride) naturally dissolved in water or at higher concentrations from the food being heated. (Have you ever noticed that the overflowing water that evaporates on the hot gas burners often leaves a white residue? That white solid is dried sodium chloride salt.) Hot sodium ions emit light of characteristic colors or frequencies, an unambiguous indicator of the presence of this element.

The Chemical Details

The element sodium is a member of the alkali metal family. Thermal ionization of alkali metals is possible in very hot flames because of the low ionization potentials of these metals. Atomic sodium, for example, has an ionization energy of 495.8 kJ mol^{-1} (≈ 5 eV).[1] Recall that the ionization energy of a neutral atom is defined as the energy required to remove the lowest-energy electron from the gaseous atom and form the positive cation: Na (g) $\rightarrow$ Na$^+$ (g) + e$^-$. The sodium ion and electron may then recombine to form a neutral but excited sodium atom. Following this thermal excitation of sodium atoms to high-energy states, sodium returns to the ground state via the emission of photons. The energy of these photons is equivalent to the wavelength of light observed. The most intense emission occurs in the yellow region of the visible range of the electromagnetic spectrum at wavelengths of 589.0 and 596.6 nm.[2]

We observe the typical yellow emission of sodium atoms in many other circumstances. Sodium vapor lamps are electric discharge lamps with metal electrodes and filled with neon gas and a small amount of sodium. Current passing through the electrodes first ionizes the neon gas. The hot neon gas then vaporizes the sodium, which is then easily excited. The characteristic yellow light emanates from the excited state species returning to the ground state. Astronomers recently discovered that one of the three tails of particles streaming from the Comet Hale–Bopp was a bright yellow tail of emitting sodium atoms.[3] Scientists at the Jet Propulsion Laboratory in Pasadena, California also reported a yellowish emission from Jupiter's moon Io arising from the cloud of sodium vapor forming a halo around Io.[4]

KEY TERMS: atomic emission spectra   ionization energy   alkali metals

References

[1] "Sodium." http://wild-turkey.mit.edu/Chemicool/elements/sodium.html

[2] "Line Spectra of the Elements," in *Handbook of Chemistry and Physics*, 74th ed., ed. David R. Lide (Boca Raton, FL: CRC Press, 1993), **10**–92.

[3] "Three-Tailed Comet." *Discover Magazine* **19** (January 1998), http://www.discover.com/cover_story/9801-2.html#3

[4] Photo Caption, Jet Propulsion Laboratory, California Institute of Technology, National Aeronautics and Space Administration, Pasadena, CA, http://nssdc.gsfc.nasa.gov/photo_gallery/caption/gal_io3_48584.txt

Related Web Sites

▶ "Chemical of the Week: Gases that Emit Light."
http://scifun.chem.wisc.edu/chemweek/gasemit/gasemit.html

Other Questions to Consider

| 8.3 | What is the dark spot left on the inside of a light bulb when it burns out? *See* p. 109.

| 12.2 | Why is an astronaut's visor so reflective? *See* p. 143.

| 15.2 | What is Liquidmetal (in Liquidmetal Golf Clubs)? *See* p. 228.

Chapter 2 ▶

Connections to Gases

GAS LAWS

2.1 What Causes an Egg to Crack if It's Boiled Too Rapidly?

Cooks often recommend that an egg be heated slowly by starting it in cold water to avoid cracking the egg upon boiling. What phenomenon, discovered in 1802, is at the heart of this advice?

The Chemical Basics

A pocket of air is formed in a hen's egg when the contents of the egg contract upon cooling after the egg is laid. This "air cell" is formed as an inner membrane separates from an outer membrane, and the air pocket is generally located at the larger end of the asymmetric egg. A close look at the shell of an egg reveals thousands[1] of tiny pores through which carbon dioxide and moisture in the egg may exit over time, allowing air to enter. As a consequence, the air pocket developed during laying increases in size as the egg ages. During the boiling of an egg, the increase in temperature causes the air pocket to increase in volume. If the boiling occurs too rapidly, the expanded air volume doesn't have time to diffuse through the porous shell, causing the egg shell to crack.

The enlargement of the air sac over time has two interesting consequences for the consumer: eggs that float in water and hard-boiled eggs with variable ease of peeling. The phenomenon of an egg floating in water is a consequence of an air cell that has enlarged sufficiently over time to keep the egg buoyant. The ease of peeling a hard-boiled egg is also related to the age of the egg. The fresher the egg, the smaller the air sac. The older the raw egg, the larger the air sac and more the egg contents must contract to accommodate the enlarged air sac. A greater degree of contraction leads to an easier-to-peel shell.

The Chemical Details

The quantitative relationship of gas volume and temperature is stated in Charles' Law:

> The volume of a gas is a linear function of its temperature.

For temperature in Celsius, the relationship is given by $V = V_0 + \alpha t$ and by $V = cT$ for absolute temperatures (Kelvin). This empirically derived relationship is commonly attributed to the French physicist, Jacques Charles (1787), and verified in 1802 by Joseph Gay-Lussac, a French chemist.

KEY TERMS: Charles' Law

References

[1] "Basic Egg Facts." American Egg Board, http://www.aeb.org/facts/facts.html#8

Related Web Sites

▶ "The Science of Boiling an Egg." Charles D. H. Williams, University of Exeter, School of Physics, http://newton.ex.ac.uk/teaching/CDHW/egg/#structure

▶ "All About Shell Eggs." Food Safety and Inspection Service, United States Department of Agriculture, August 1999, http://www.fsis.usda.gov/OA/pubs/shelleggs.htm

▶ "Charles Law." John L. Park, 1996, http://dbhs.wvusd.k12.ca.us/GasLaw/Gas-Charles.html

▶ Joseph-Louis Gay-Lussac, Excerpts from "The Expansion of Gases by Heat." *Annals de Chimie* **43**, 187 (1802), http://webserver.lemoyne.edu/faculty/giunta/gaygas.html

Other Questions to Consider

5.19   What causes puff pastry to expand? *See* p. 67.

VAPOR PRESSURE & VOLATILITY

2.2 # When a Flambé Is Prepared, Why Is the Liqueur Heated Prior to Lighting the Flame?

Cherries Jubilee, Bananas Foster, Crèpes Suzette, Beef Flambé, Steak au Poivre—all are recipes that use a dramatic procedure of flaming a liqueur to introduce additional flavors into the preparation. A little knowledge of chemistry will ensure that your concoction is a delicious one!

The Chemical Basics

Many recipes call for a *warmed* liqueur to be ignited with a match. This procedure, known as *flambé* or to flame, enhances the flavor of a dish through the process of carmelization. The ignition of the flame evaporates the alcohol and helps the flavor of the liqueur to blend into the food. Since the liqueur is to be ignited with a flame, you might wonder why the recipe instructs you to warm the liqueur prior to flaming. The liquid itself is not the substance that sustains the flame. The vapors associated with the alcohol component of the liqueur are responsible for the spectacular appearance. Heat facilitates flaming by producing vapors of the alcohol which ignite more easily than the liquid phase.

The Chemical Details

A fire will not occur until a *flammable* (i.e., capable of being ignited) liquid is heated above a certain temperature called the *flash point*. The flash point of a liquid is the lowest temperature at which the liquid gives off enough vapor to ignite on exposure to a flame. A cold liqueur, that is, liquid at a temperature below its flash point, does not produce enough vapors to burn. (The *vapor pressure* of a liquid increases with temperature; i.e., the amount of vapor in equilibrium with the liquid form of the substance increases as temperature is raised.) The flammable ingredient in liqueurs is ethanol with a flash point of 55°F. Cold or even room-temperature rums, brandies, and liqueurs will not support a flame for two reasons. The first reason is that the concentration of alcohol in the water is too low for the water/alcohol mixture to burn. A 50-50 mixture of ethanol and water has a flash point of 75°F. The second reason, actually related to the first, is that the vapor pressure of the flammable alcohol over the liqueur solution is too low to be ignited. The concept of proof in alcoholic beverages relates to the amount of alcohol in the mixture. Alcoholic proof liquors less than 100 proof can not burn. Most liqueuers are less than 100 proof. However, when a liqueur-containing mixture of less than 100 proof is heated, vapors will be readily formed above the surface of the skillet or pan that will ignite if a match is brought close to the surface.

2.3 How Do Correction Fluids like Liquid Paper and White-Out Work?

Typewriter correction fluid was first invented by Bette Nesmith Graham.[1,2] Her early version for the product consisted of a white tempura paint to correct mistakes. Looking for a mixture that was both quick-drying and barely detectable, she discovered a formulation that she would later sell to the Gillette Company in 1979. The formula was modified slightly in the 1980s. What is the chemistry of Liquid Paper?

The Chemical Basics

Paper correction fluids contain two key ingredients — a white (or colored) pigment and a volatile fluid solvent. The pigment is initially dissolved in the solvent, but, upon application to a surface, the solvent readily evaporates. A solid residue of pigment remains. If solvent evaporates from an open bottle of correction fluid, additional solvent in the form of correction fluid thinner can be added to redissolve the solid pigment.

The Chemical Details

White-colored correction fluids like Liquid Paper and White-Out contain the pigment titanium dioxide (TiO_2) and the volatile solvent 1,1,1-trichloroethane or methyl chloroform (CCl_3CH_3).[3] The volatile substances in the correction fluid contribute 50% of the total volume of the product.[3] Correction fluid thinner is simply 100% 1,1,1-trichloroethane solvent,[4] added to redissolve any solidified titanium dioxide.

References

[1] Bette Nesmith Graham (1924–1980): Liquid Paper, Inventor of the Week Archives, The Lemelson-MIT Prize Program,
http://web.mit.edu/invent/www/inventorsA-H/nesmith.html

[2] Who was the Inventor of Liquid Paper™, PageWise, Inc.,
http://tntn.essortment.com/inventorliquid_ridn.htm

[3] Liquid Paper Correction Fluid, White; Material Safety Data Sheet,
     http://www.biosci.ohio-state.edu/~jsmith/MSDS/LIQUID%
     20PAPER%20CORRECTION%20FLUID%20WHITE.htm

[4] Liquid Paper Correction Fluid Thinner, MSDS Sheet,
     http://www.biosci.ohio-state.edu/~jsmith/MSDS/LIQUID%
     20PAPER%20CORRECTION%20FLUID%20THINNER.htm

2.4   Why Do Cosmetic Cold Creams Feel Cool When Applied to the Skin?

The terminology "cold cream" aptly describes the soothing and cool-ing sensation of these cosmetics. Some basic chemistry fundamen-tals explain this highly marketed phenomenon.

The Chemical Basics

The cooling sensation experienced after applying a cosmetic cold cream on one's skin is the result of the evaporation of an alcohol (e.g., ethanol) contained in the cold cream. Formulators of skin products include ethanol to achieve a variety of benefits. For example, alcohol enhances the ability of the components in the cold cream to dissolve. For the consumer, the presence of alcohol eases the applica-tion of the cream on the skin, enhances the perfume quality of the mixture, and provides a cooling effect on the skin.

Why does the evaporation of an alcohol achieve a cooling effect? Evaporation is a process that requires heat (an endothermic process). The skin in the area of application provides the necessary amount of heat needed to evaporate the alco-hol, leading to the localized cooling sensation.

The Chemical Details

While the particular ingredients of personal care products are often proprietary, one manufacturer, Uniqema, has provided the formulations for two of its "hy-droalcoholic skin care emulsions."[1] In each of these products, ethanol is present at a level of 20.00% by weight. Chesebrough-Pond's, Inc., in the patent abstract for a cold cream cosmetic composition,[2] indicates that a *polyhydric alcohol* (an alcohol with more than one hydroxyl group, –OH) with two to six carbons is the volatile species that provides the cooling sensation. The relatively low boiling points and correspondingly high vapor pressures of substances like ethanol and polyhydric alcohols are factors in promoting the ease with which these materi-als evaporate and provide a cooling sensation. For example, the vapor pressure of ethanol is 59 torr at 298 K (25°C).[3] How does this value compare with that for water, a substance that we recognize also evaporates easily from one's skin? Water has a vapor pressure of 24 torr at the same temperature;[4] thus ethanol vaporizes even much more readily than water. (Recall that the higher the vapor pressure of a liquid at a fixed temperature, the greater the tendency to escape to the gas phase.)

KEY TERMS:	vapor pressure	volatility	evaporation
	vaporization	endothermic	

References

[1] "Personal Care: Hydroalcoholic Skin Care Emulsions." Uniqema, http://surfactants.net/formulary/uniqema/pcm14.html

[2] "Cosmetic Compositions including Polyisobutene." U.S. Patent 5695772, December 9, 1997.

[3] "Properties of Common Solvents," in *Handbook of Chemistry and Physics*, 74th ed., ed. David R. Lide (Boca Raton, FL: CRC Press, 1993), **15**–46.

[4] "Vapor Pressure of Water from 0 to 370°," in *Handbook of Chemistry and Physics*, 74th ed., ed. David R. Lide (Boca Raton, FL: CRC Press, 1993), **6**–15.

Other Questions to Consider

5.19   What causes puff pastry to expand? *See* p. 67.

7.1   Why are general anesthetics administered as gases? *See* p. 78.

13.6   How do "sniffing dogs" detect narcotics or explosives? *See* p. 176.

Chapter 3 ▶

Connections to Solutions

Questions to Consider

| 10.1 | Why do CAT scans often cause a "warm flush" sensation? *See* p. 128.

SOLUBILITY

| 3.1 | ## Why Do Carbonated Drinks Go Flat as They Warm?

The ideal serving temperature of a carbonated beverage is often stated to be near 40°F to maintain a high degree of carbonation. Many variables can affect how quickly the bubbliness of the beverage is lost, but temperature is a critical factor. The behavior of gases and liquids as a function of temperature plays an important role in how refreshing we find our favorite carbonated beverage.

The Chemical Basics

While carbonated beverages represent a multibillion dollar industry in modern society, the carbonation process for producing soft drinks was first developed in the United States in the 1770s. What prompted both European and American scientists to search for viable methods to produce carbonation in beverages? The desire was attributed to the reputed healthful properties of the naturally effervescent mineral waters found at various springs throughout Europe. Today carbonation — in the form of carbon dioxide (CO_2) gas — is introduced in beverages to add taste and effervescence to appeal to both the tastebuds and the eye. The added CO_2 bubbles also enhance shelf life and prevent spoilage. The carbonation process takes advantage of the particular conditions — cold temperatures and

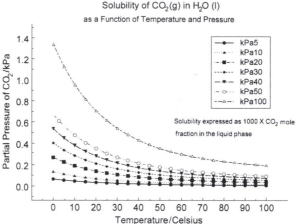

Figure 3.1.1 ▶ The solubility of gaseous carbon dioxide in water as a function of both temperature and pressure. The CO_2 solubility is expressed in terms of the mole fraction of carbon dioxide in the liquid solution.

high pressures — that favor the mixing of a gas and a liquid such as water. To introduce the CO_2 gas into the water, the liquid beverage mixture is chilled below 40°F and then cascaded over a series of plates in an enclosure containing carbon dioxide gas under pressure (3–4 times higher than atmospheric pressure). The amount of gas that the water will absorb during this process is enhanced by the cold temperatures and the high pressures. Whenever a bottle or can of carbonated beverage is opened, the decrease in pressure lowers the ability of the CO_2 gas to stay dissolved in the water solution. As the temperature of the beverage rises, the solubility of the dissolved CO_2 gas also decreases. The loss of effervescence is evidence of this decrease in gas solubility. This behavior is directly opposite to our experience with the solubility of solids in water. Most solids (for example, sugar or salt) dissolve more readily in hot water than in cold.

The Chemical Details

The solubility of nearly all gases in water decreases as the temperature is increased. Furthermore, the solubility of a gas increases with the partial pressure of the gas above the surface of a liquid solution, expressed as *Henry's Law*:

$$S_{\text{gas}} \text{ (moles/liter)} = k_H \times P_{\text{gas}} \text{ (atm)}.$$

Both of these facts are employed in the carbonation process of sodas and beer and some sparkling wines. Low-temperature conditions and CO_2 pressures of 3 to 4 atm are used to enhance the dissolution of carbon dioxide gas in water. The graph in Fig. 3.1.1 presents the solubility of carbon dioxide in water at various temperatures and pressures.[1] The parameter used to express CO_2 solubility is the mole fraction of CO_2 in the liquid phase. As a further note, the added tang of carbonated beverages results from the formation of the weak acid carbonic acid in water, defined by the equilibrium reaction

$$CO_2 \text{ (g)} + H_2O \text{ (l)} \rightleftharpoons H_2CO_3 \text{ (aq)} \rightleftharpoons H^+ \text{ (aq)} + HCO_3^- \text{ (aq)}.$$

KEY TERMS:　solubility　Henry's Law　partial pressure

References

[1] "Solubility of Carbon Dioxide in Water at Various Temperatures and Pressures," in *Handbook of Chemistry and Physics*, 74th ed., ed. David R. Lide (Boca Raton, FL, CRC Press: 1993), **6**–7.

3.2　Why Do Crystals Form in Wine over Time?

A wine bottle with sediment or crystals at the bottom is not flawed. While many consumers find such crystals to be unattractive, vintners know that crystal formation is commonplace in wine production. What chemistry explains the origin of the crystals in wine?

The Chemical Basics

The complexity of wine composition is a central reason for the vast variety of wines in the marketplace. In addition to water and ethanol, the major components, a variety of organic acids as well as metal ions from minerals in the skin of the grape are present. Initially, all of these substances remain dissolved in the bottled grape juice. As the fermentation process occurs, the increasing alcohol concentration in the wine alters the solubility of particular combinations of acid and metal ions. Unable to remain in solution, the insoluble substances settle as crystals. Since the process of red-wine making involves extended contact of the grape juice with the skins of the grapes (where the minerals are concentrated), wine crystals are more common in red wines than in white wines.

The Chemical Details

Grapes are one of the few fruit crops that contain a significant amount of the weak organic acid known as *tartaric acid*, $HOOC-(CHOH)_2-COOH$. More than half of the acid content of wine is ascribed to tartaric acid. As a weak acid, tartaric acid partially ionizes in water to yield the bitartrate or hydrogen tartrate ion:

$$HOOC-(CHOH)_2-COOH(aq) \rightleftharpoons H^+ \text{ (aq)} + HOOC-(CHOH)_2-COO^-(aq).$$

The bitartrate ion can combine with potassium ion, also present in high concentrations in grapes, to form the soluble salt potassium bitartrate (also known as cream of tartar). In water sodium bitartrate is fairly soluble: 1 g dissolves in 162 ml of water at room temperature.[1] In alcohol solution (formed as fermentation of the wine yields ethanol), the solubility of potassium bitartrate is significantly reduced: 8820 ml of ethanol are required to dissolve 1 g of the salt.[1] As a consequence deposits of potassium bitartrate form as the salt precipitates out of solution.

To prevent the formation of wine crystals during the bottling process, winemakers use a method known as *cold stabilization*. By lowering the temperature of the wine to 19–23°F for several days or weeks, the solubility of tartrate crystals is lowered, forcing the crystals to sediment. The resulting wine is then filtered off the tartrate deposit. The temperature dependence of the solubility of potassium bitartrate is readily apparent in the following comparison: while 162 ml of water at room temperature dissolves 1 g of the salt, only 16 ml of water at 100°C are needed to solubilize the same amount of salt.[1] Recent developments employ a technique known as electrodialysis to remove tartrate, bitartrate, and potassium ions from newly fermented wine at the winery before potassium bitartrate crystals form.[2]

KEY TERMS: solubility

References

[1] M. Windholz and S. Budavari, eds., *The Merck Index: An Encyclopedia of Chemicals, Drugs, and Biologicals*, 10th ed. (Rahway, NJ, Merck: 1983).

[2] Steve Heimoff, "Electrodialysis Removes Crystals But Technology Has Been Ahead of Its Practicability." *Wine Business Monthly*, May 1996, http://winebusiness.com/html/SiteFrameSet.cfm?fn=../Archives/Monthly/1996/9605/bm059612.htm

3.3 What Makes a Lava Lamp Work?

Many fundamental chemical principles underlie the operation of this intriguing novelty device.

The Chemical Basics

The lava lamp has captivated viewers for over fifty years. Shortly after discovering the lava lamp's prototype — a "contraption made out of a cocktail shaker, old tins, and things" — in a Hampshire, England pub after World War II, Edward Craven Walker founded the Crestworth Company in Dorset, England to develop the device.[1] Over a 15-year period he perfected the lava lamp and mass marketed the light source. While the demand for lava lamps reached fad proportions in the 1960s, a resurgence of the lava lamp craze occurred in the 1990s. Walker's company, sold in 1990, still produces lava lamps, and Haggerty Enterprises of Chicago produces Lava Lites in the United States. A vast number of color combinations exist.

Two liquids with vastly different properties are the essential ingredients of a lava lamp. The two liquids must differ significantly in chemical composition to prevent mixing. The liquid that constitutes the "lava" is generally a substance that does not dissolve readily in water (i.e., a nonwater soluble substance). The second liquid is a water-soluble material or a mixture of water and a liquid that dissolves readily in water. The inability of the "lava" and the water-based solution to mix encourages the lava to coalesce to form the familiar "blobs" that migrate through the water-based liquid. The lava material must be slightly more dense than water at room temperature, positioning the lava at the bottom of the container before the lamp is switched on. An illuminated lamp, usually a 40-W bulb, serves as a heat source. A small coil of wire at the base of the lamp also serves as a heating element within the liquid compartment. Both liquids decrease in density as the temperature of the system increases, but the liquids are selected so that the lava liquid changes density more dramatically than the water solution as the temperature rises. The now less dense ("lighter") lava rises in the container and also expands in volume due to higher temperatures. A liquid with a moderate viscosity contributes to the motion and circulation of the lava. As the lava moves away from the heat source, it begins to cool, decreasing its volume (contracting the "blobs") and increasing its density (sinking to the bottom of the lamp). The cycle repeats to generate the mesmerizing and ever-changing pattern formation and motion of the lava. Different colored dyes — one that dissolves readily in a water-based mixture and a second that disperses only in the lava liquid — are added to produce the various color combinations.

The Chemical Details

Liquids with disparate intermolecular forces are critical for the operation of a lava lamp. The patent issued in 1971[1] specifies mineral oil (a hydrocarbon-based liquid) and a 70/30% by volume water/propylene glycol mix as the insoluble liquids. Other water-insoluble substances with specific gravities greater than that of water (i.e., densities greater than water) and with larger thermal coefficients of expansion (i.e., greater temperture-dependent densities to increase liquid volume upon heating) include benzyl alcohol, cinnamyl alcohol, diethyl phthalate, and ethyl salicylate. Isopropyl alcohol, ethylene glycol, and glycerol are water-soluble substitutes for propylene glycol. The hydrophobic forces that dominate in the lava liquids are incompatible with the hydrogen bonds and dipole interactions that exist in the water-based mixtures. Since the forces of attraction to produce a single liquid phase are not present, two distinct phases emerge.

Six patents that reference Cravens original patent have been issued. One of the earlier patents[3] details a three-phase liquid system to create a "device for producing aesthetic effects." The patent specifications involve at least two liquids of different density that are not completely or permanently miscible with each other. In particular the three liquids A, B, C, denoted in order of increasing density, are described as: "A liquid paraffin and/or silicone oil and/or naphthene and/or hexachlorobutadiene; B water or an ether, more particularly propanetrioxyethylether or a polyether; C esters with chemically bound phosphorus and/or chemically bound halogen, preferably chlorine, an ester of phthalic acid, and more particularly dibutoxybutyl phthalate, a carbonic acid ester, more particularly propanediol carbonate, or ethanediolmonophenylether or tetrahydrothiophene-1,1-dioxide with the provision that the selected liquids are not completely and not permanently miscible with each other."[4] In a later patent (US4419283: 12/06/1983, "Liquid Compositions for Display Devices"), an invention consisting of systems of three, four, and five mutually immiscible liquid phases suitable for use in display devices and novelty toys is described. The preferred four-phase systems comprised one highly hydrophobic organic phase, one organic phase containing compounds that are moderately polar, one phase containing hydrogen-bonding organic compounds, and one aqueous phase. Numerous examples of specific liquids that fulfill these conditions are delineated. As long as there is interest in such novelty devices as lava lamps, the search continues for room-temperature, mutually immiscible liquid systems composed of water and organic liquids that are inexpensive, nontoxic, noncombustible, and readily dyeable!

KEY TERMS: miscibility   solubility   density

References

[1] "Lava Lite Lamp." [US Patent 3,570,156, March 16, 1971], Your Mining Co. Guide to Inventors; http://inventors.miningco.com/library/weekly/aa092297.htm

[2] R. Hubscher, "How to Make a Lava Lamp." *Popular Electronics* (March 1991), 31: also in *Popular Electronics, 1992 Electronics Hobbyist's Handbook.*

[3] These patents are U.S. Patent 5778576, issued on 07/14/1998, "Novelty lamp"; 5709454, 01/20/1998, "Vehicle visual display devices"; D339295, 09/14/1993, "Combined bottle and cap"; 4419283, 12/06/1983, "Liquid compositions for display devices"; 4085533, 04/25/1978, "Device for producing aesthetic effects"; 4034493, 07/12/1977, "Fluid novelty device."

[4] "4085533: Device for producing aesthetic effects." IBM US Patent Database, Patent 4085533.

[5] "4419283: Liquid compositions for display devices." IBM US Patent Database, Patent 4419283.

Related Web Sites
▶ "Self-Made Lavalamps." Robert Fenk, http://www.robf.de/Lavalamps/

| 3.4 | # How Is the Toxicity of Barium Sulfate Controlled for X-Rays? |

Barium sulfate is a radiopaque or contrast agent that blocks the transmission X-rays to help physicians see any special conditions that may exist in an organ or that part of the body where the contrast agent is localized. The areas in which radiopaque agents are located will appear white on X-ray film, creating the needed distinction, or contrast, between one organ and other tissues. How can the toxicity of the barium ion be controlled to provide this necessary radiopaque function?

The Chemical Basics

Body tissues absorb X-rays in differing degrees. The greater the extent of absorption of the X-ray light, the lighter the tissue will appear on the radiograph, enabling a physician to monitor the condition of the tissue. To enhance the ability to "see" the gastrointestinal tract, a barium sulfate suspension is ingested before X-rays are administered to trace a person's digestive tract. Barium sulfate absorbs (i.e., does not transmit) X-rays, and thus barium sulfate provides a radiopaque contrasting medium. In other words, with the added barium sulfate, the lining of the stomach and the gastrointestinal tract will be coated with barium sulfate and will appear white on an X-ray. Although all soluble barium salts are toxic, the procedure is safe because $BaSO_4$ is essentially insoluble. To limit the ability of barium sulfate to dissolve in solution and release toxic barium ions, some sodium sulfate is added to the suspension. Sodium sulfate is exceedingly soluble in water, releasing sulfate anions. The elevated concentration of sulfate ions in solution restricts the dissolution of barium sulfate, as such a dissolution would generate more sulfate anions. Hence, the barium sulfate remains essentially intact, limiting the concentration of barium ions in the body.

The Chemical Details

The dissolution of barium sulfate in water is described by the reaction

$$BaSO_4 \text{ (aq)} \rightleftharpoons Ba^{2+} \text{ (aq)} + SO_4^{2-} \text{ (aq)}.$$

The limited solubility (S, moles per liter) of barium sulfate is reflected in the solubility product for the dissolution. The solubility can be calculated to show that a small amount of barium ion is found free in solution:

$$K_{sp} = 1.1 \times 10^{-10} = S^2$$
$$S = [Ba^{2+}] = 1.05 \times 10^{-5} \text{ M}.$$

Barium sulfate suspensions also contain 0.10 M Na_2SO_4 and use the common ion effect to control the solubility of barium sulfate in solution. Since Ba^{2+} is toxic to humans, we want to adjust conditions to ensure that the direction of the dissolution reaction favors the reactants. The salt Na_2SO_4 will readily dissociate into Na^+ and SO_4^{2-} ions and thus increase the concentration of SO_4^{2-}. By Le Chatelier's principle, the system will respond to the presence of the common ion (i.e., the ion present in both the soluble Na_2SO_4 and the sparingly soluble $BaSO_4$) by shifting the position of the original dissolution equilibrium to the left. In other words, the solubility of the reactant barium sulfate will be further decreased by the presence of the sulfate anion. What is the solubility (S' in moles per liter) of barium sulfate in the presence of 0.10 M Na_2SO_4? Using the same approach as before,

$$K_{sp} = 1.1 \times 10^{-10} = (S')(S' + 0.10)$$
$$S' = [Ba^{2+}] = 1.1 \times 10^{-9} \text{ M}.$$

Thus, by an application of the common ion effect, the solubility of barium sulfate has been reduced to produce 10^4-fold less free barium ion in solution, thus further reducing the risk of barium toxicity.

KEY TERMS: solubility   dissolution   common ion effect
                      Le Chatelier's principle

Related Web Sites

▶ "Interpreting the Radiograph." Dr. Simon Cool,
    http://www.uq.edu.au/~anscool/xray/xray.html

▶ Barium Sulfate (Diagnostic), MEDLINEplus Health Information, National Institutes of Health, http://www.nlm.nih.gov/medlineplus/druginfo/bariumsulfatediagnostic 203643.html

3.5   Why Is Milk of Magnesia an Antacid?

Magnesium-containing compounds are often common everyday medicines. What makes milk of magnesia an effective antacid?

The Chemical Basics

In pharmacology, the term "milk" refers to an aqueous suspension of a water-insoluble drug. The antacid milk of magnesia contains a saturated solution of the salt magnesium hydroxide. Magnesium hydroxide does not dissolve readily in water, and a saturated solution this substance simply means that so much magnesium hydroxide has been added to water such that not all of the solid will dissolve. Instead of the nondissolving solid settling to the bottom of the container, the solid particles remain suspended in the solution, hence the characterization as a "milk." The small amount of magnesium hydroxide that does dissolve in water yields an alkaline component (hydroxide ion) that counteracts the acid in the stomach.

The Chemical Details

The solubility S of magnesium hydroxide (in units of moles per liter) can be computed from the equilibrium constant (the *solubility product*) for the dissolution of the salt. For S equal to the moles per liter of $Mg(OH)_2$ dissolved:

$$Mg(OH)_2 \ (s) \ \rightleftarrows \ Mg^{2+}(aq) + 2OH-(aq), \ K_{sp} = 1.2 \times 10^{-11}$$
$$K_{sp} = [Mg^{2+}][OH^-]^2 = 1.2 \times 10^{-11}$$
$$S(2S)^2 = 1.2 \times 10^{-11}$$
$$4S^3 = 1.2 \times 10^{-11}$$
$$S = 1.44 \times 10^{-4} \ M.$$

With a molecular weight for magnesium hydroxide of 58.3 g/mole, the solubility is equivalent to 8.36 mg per liter. The Bayer product "Original Phillips' Milk of Magnesia" contains 400 mg of magnesium hydroxide per teaspoon (volume estimated at 5 mL)[1] or 8.00×10^4 mg per liter. Clearly a saturated solution! From the amount of dissolved $Mg(OH)_2$, the pH of the solution can be determined, for

$$[OH^-] = 2S = 2.9 \times 10^{-4} \ M$$
$$[H_3O^+] = \frac{K_w}{[OH^-]} = \frac{1.00 \times 10^{-14}}{2.9 \times 10^{-4}} = 3.5 \times 10^{-11} \ M$$
$$pH = -\log_{10}[H_3O^+] = 10.46.$$

The alkaline nature of the saturated $Mg(OH)_2$ solution provides the antacid capabilities of this over-the-counter medicine.

KEY TERMS: solubility alkaline saturated solution

References

[1] The Phillips Medicine Cabinet, Bayer Corporation,
http://www.bayercare.com/htm/philmom.htm#ingredients

Other Questions to Consider

| 5.16 | How does washing soda soften hard water? *See* p. 61.

| 7.2 | Why won't water relieve the burning sensation of chili peppers? *See* p. 83.

| 10.5 | What causes an instant ice-pack to cool? *See* p. 134.

COLLIGATIVE PROPERTIES

| 3.6 | # Why Are Ice Cubes Cloudy on the Inside?

Pure, clear liquid water should freeze as a clear solid, but ice cubes commonly have a cloudy appearance. What chemistry should you know to recognize that cloudy ice cubes are not cause for serious alarm?

The Chemical Basics

Water is rarely a "pure" substance but contains both dissolved gases (e.g., oxygen) from the atmosphere and dissolved minerals (e.g., calcium and magnesium salts). The presence of these substances affects the temperature at which water freezes. Pure water freezes at $0°C$; water with dissolved gases and mineral salts freezes at a lower temperature. The higher the concentration of dissolved gases and minerals, the lower the freezing point of water. As water cools, the first layer of ice that forms is at the interface with air. As ice forms, *pure water* solidifies, leaving the dissolved gases and salts in solution. Thus, the freezing process concentrates the dissolved species in smaller and smaller volumes of liquid solution, effectively increasing their concentration. With a higher concentration of dissolved material, the temperature at which additional ice will form is lowered. The cloudiness in the center of an ice cube thus is the consequence of the concentration of dissolved gases and minerals that refract light and create an opaque appearance.

The Chemical Details

The depression of the freezing point of a solvent due to the presence of a dissolved solute is an example of a *colligative property*, that is, a property of a dilute solution that depends on the number of dissolved particles and not on the identity of the particles. Water has a *freezing point depression constant*, K_f, of 1.86 K kg mol^{-1}. In other words, for every mole of nonvolatile solute dissolved in a kilogram of water, the freezing point of water is lowered by $1.86°C$. The change in freezing point, ΔT, can be calculated from the equation

$$\Delta T = -K_f\, m,$$

where m is the molality of the solution, the number of moles of solute dissolved per kilogram of solvent (water). If the solute dissociates into ions upon dissolution in water, then m must be expressed as the *total molality* of all species (nondissociating or ionic) in the solution. Thus, one mole of NaCl in a kilogram of water will lower the freezing point by $2 \times 1.86°C$ as a consequence of the two moles of ions (Na^+ and Cl^-) present in solution. In contrast, one mole of sugar in one kilogram of water would lower the freezing point by $1.86°C$, for sugar does not dissociate into ions when dissolved.

KEY TERMS:	freezing point depression	molality	colligative property
	solvent	solute	

Related Web Sites

► "On the Theory of Electrolytes. I. Freezing Point Depression and Related Phenomena. (Zur Theorie der Elektrolyte. I. Gefrierpunktserniedrigung und verwandte Erscheinungen)." P. Debye and E. Hückel (Submitted February 27, 1923), Translated from *Physikalische Zeitschrift*, Vol. 24, No. 9, 1923, pages 185–206, Classic Papers from the History of Chemistry,
http://dbhs.wvusd.k12.ca.us/Chem-History/Debye-Strong-Electrolyte.html

COLLOIDAL DISPERSIONS

3.7   Why Are Metal or Glass Bowls Best for Whipping Egg Whites?

Recipes for meringues often include instructions such as: "In a large glass or metal mixing bowl, beat egg whites and cream of tartar until soft peaks form. Gradually beat in sugar then vanilla until stiff peaks form." Why do professional bakers and experienced cooks (as well as knowledgeable chemists!) heed the recipes' call for a metal or glass mixing bowl?

The Chemical Basics

Vigorous mixing of egg whites introduces an extensive amount of air bubbles, producing a foam that retains its structure during baking. A voluminous amount of whipped egg whites is achieved by ensuring that no fat is present. Egg yolks contain fats or lipids, so the separation of yolks from whites is essential. Plastic bowls and utensils are porous and often retain fats even after washing. The surfaces of glass and metal bowls are essentially fat free, producing the most extensive degree of whipping.

The Chemical Details

A foam is a *colloidal dispersion* of gas bubbles trapped in a liquid. To produce a stable foam, several characteristics of the liquid are necessary. For example, a viscous liquid facilitates the trapping of gas bubbles. The presence of a surface active agent or stabilizer that, for structural reasons, preferentially locates on the surface of the gas bubble also provides a more permanent foam. A low vapor pressure for the liquid reduces the likelihood that the liquid molecules (particularly those surrounding the bubble) will easily evaporate, thus leading to the collapse of the foam.

The albumen of egg white is a protein solution that foams readily upon whipping. Research suggests that the proteins ovomucin, ovoglobulins, and conalbumins are primarily responsible for foam formation. The proteins collect at the air—water interface of the air bubble and denature (unfold) to support the foam structure. Further denaturation through heating (baking) coagulates the proteins to result in a more stable structure. The addition of sugar during beating enhances the formation of the egg white foam due to the hygroscopic nature of sugar that retains water. (The hydroxyl groups on the sugar structure form hydrogen bonds with water.) However, sugar retards denaturation, and therefore more beating is required to reach the same extent of foam formation, particularly if sugar is added too early in the whipping process. The addition of cream of tartar, an acid (tartaric acid), lowers the pH of the protein solution, facilitating protein denaturation and coagulation. Fat, if present, would also tend to collect at the air—water interface of the bubble. However, fat, unlike protein, does not denature but coagulates. Thus, the presence of fat reduces the ability of the protein to denature and stabilize the foam.

KEY TERMS: foam   colloidal dispersion

Related Web Sites

► "Eggs and Egg Products." Science of Foods, Oregon State University,
   http://food.orst.edu/nfm236/egg/index.cfm

► "Food Study Manual." C. Daem and D. Peabody, The School of Family & Nutritional Sciences, University of British Columbia,
   http://www.library.ubc.ca/ereserve/hunu201/fdmanual/

3.8 How Do Fog Machines Create the Artificial Fog or Smoke Used in Theatrical Productions?

Whether for a dramatic scene or a science-fiction fantasy, the use of artificial fog to create ambience and special effects is commonplace in theatrical and film productions. On what basic chemical principles do fog machines operate?

The Chemical Basics

Some of the most memorable scenes in a motion picture are the creation of special effects technicians. One standard tool in the entertainment industry to create ambience, simulate specific designs such as volcanoes or swamps, or accentuate optical effects is fog production. (In the theatrical sense, the terms *fog* and *smoke* are used interchangeably for mist consisting entirely of liquid droplets. In the chemical sense, fog refers to a liquid phase dispersed in a gas, while *smoke* contains solid particulate matter dispersed in a gas.) There are a variety of approaches to creating artificial fog or smoke.

Many companies sell machines that use a glycol- and water-based fog fluid.[1,2] These liquids are pumped into a temperature-controlled heat exchanger, heated, and, as a consequence, vaporized into thick clouds of fog. Different percentages of propylene glycol (Fig. 3.8.1) and triethylene glycol (Fig. 3.8.2) help to achieve

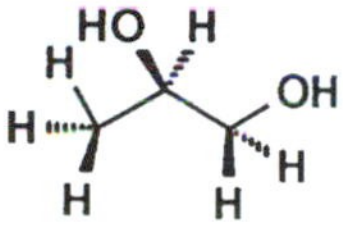

Figure 3.8.2 ▶ The molecular structure of propylene glycol, used in artificial fog machines.

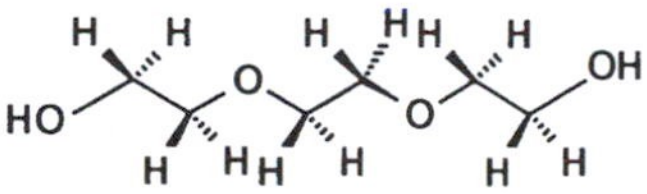

Figure 3.8.3 ▶ The molecular structure of triethylene glycol, used in artificial fog machines.

the desired effect, for example, by creating an invisible particulate mist that enhances the viewing of laser light beams or by concocting a dense, white fog with varying "hang times." These thermally evaporated glycol-based conventional fog machines are preferred over "cracked-oil" machines that rely on atomization produced by pressurization.[3] Both enhanced fire safety and reduced residue deposits despite faster dissipation times are characteristic of the thermal evaporation machines.

A number of health concerns have been raised with glycol-based fog machines.[4] In particular, the *hygroscopic* nature (i.e., tendency to absorb moisture) of glycols is purported to cause a drying effect on the nose, eyes, and throat. Some manufacturers have designed methods for creating superfine fog droplets solely from water without excessive wetting. The Academy of Motion Picture Arts and Sciences Board of Governors granted a Technical Achievement Award in 1998 to three employees of Praxair, Inc., for their creation of an artificial fog that uses a mixture of Praxair oxygen, nitrogen, and steam.[5] Praxair also provided a special effects mixture of oxygen and nitrogen during the filming of Columbia Pictures' Spider-Man in 2002.[6]

How can one keep the fog low to the ground? Chilling the fog will help. By lowering the temperature of the fog, the density of the fog will increase and the seemingly heavier mist will sink to lower levels. Some foggers utilize dry ice or liquid CO_2 to produce a low-lying fog effect.[7] These refrigerants are used to cool a manifold through which the fog passes.

As an interesting application of fog-making technology, companies have designed *fog security systems*. These security systems generate a dense fog to reduce visibility to zero and trap would-be thieves in the fog until authorities arrive on the scene.[8,9]

The Chemical Details

Why are mixtures of *polyfunctional* alcohols (i.e., containing more than one hydroxyl –OH group) used to produce smoke and fog effects? As we've seen, small liquid particles dispersed in a gas create a fog. Initially, the mixture of propylene glycol, triethylene glycol, and water is liquid. The resulting fog created by vaporizing a mixture of these liquids is dense, white, and odorless. In actuality, primarily the water in these mixtures is vaporized (boiling point of 100°C), while propylene glycol (boiling point of 187.6°C) and triethylene glycol (boiling point of 285°C) remain largely in the liquid phase. The *hygroscopic* nature of propylene glycol and triethylene glycol arises from their ability to form strong hydrogen bonds and ensures the favorable dispersion of small droplets of these liquids in the water vapor. The white color of fog, just like the white color of clouds, is a consequence of the equal scattering of all wavelengths of light by the suspended liquid droplets.

KEY TERMS:	fog	smoke	hygroscopic
	hydrogen bonding	polyfunctional alcohol	

References

[1] "Atmospheres Lighting Enhancement Fluid." HighEnd Systems, Lightwave Research, http://www.highend.com/support/product/ps_f100_msds.htm

[2] "A Fog & Haze Machine Primer." FogFactory.com, http://www.fogfactory.com/index.htm

[3] "Oil-Cracking Fog Machines." Dennis Griesser, http://wolfstone.halloweenhost.com/TechBase/fogoil_OilCracker.html

[4] "Glycol, Glycerin, and Oil Fog Reading List." Entertainment Services & Technology Association, www.esta.org/tsp/fogdocs.htm

[5] "Holy Fog Bank, Batman!," Industry News, Micro Magazine.com, April 1998, http://www.micromagazine.com/archive/98/04/lefty.html

[6] Praxair's SaFog Mixture Featured in SpiderMan (2002), Praxair, http://www.praxair.com/praxair.nsf/AllContent/9ACC5354C3BFA75B85256 BA7005D5B47?OpenDocument

[7] "Coldflow LCO2 Exchanger Module." Intelligent Lighting Creations, http://www.intelligentlighting.com/Coldflow.htm

[8] "Smokecloak: Information: The Concept." Smokecloak Ltd., http://www.smokecloak.com/

[9] "Development Concept Behind Fog Security Systems." Fog Security Systems Inc., http://www.fogsecurity.com/concept/index.html

Related Web Sites

▶ "How Fog Machines Work." Halloween OnLine, http://www.gotfog.com/fog_machine_how.html

3.9 Why Are Opals and Pearls Iridescent?

For centuries the gemstone opal has been admired for its captivating and distinctive play of color. The name of this gemstone is attributed to various origins, including the Latin word *opalus* and the Sanskrit word *upala*, both meaning precious stone or jewel, and the Greek term *opallios*, translating as "to see a change of color." Although revered as "the queen of gems" in Shakespeare's *Twelfth Night*, the opal suffered in popularity in the nineteenth century when Sir Walter Scott associated the opal with bad luck in the novel *Anne of Geierstein*. What is the chemical origin of the opal's dramatic rainbow of flashes of colors?

The Chemical Basics

A variety of minerals are prized for their exquisite beauty, rarity, and exceptional durability. These extraordinary materials are classified as gemstones. One such mineral, silica, with a chemical composition of SiO_2 (silicon dioxide), exhibits several crystal structures. Several gemstones are crystalline forms of silica, including amethyst, aquamarine, emerald, garnet, peridot, topaz, tourmaline, and zircon.[1]

The gem opal is an essentially noncrystalline or amorphous form of silica that contains one or two silica minerals. Opal's gemstone quality is primarily derived from its *iridescence* — the interference of light at the surface or in the interior of a material that produces a series of colors as the angle of incidence changes. A very common example of this phenomenon is the observation of color fringes or bands when looking at a thin layer of spilled gasoline. What is the source of this iridescence in opal? A combination of interference and diffraction effects result from the structure of opal. Opal is an example of a *solid emulsion* — a heterogeneous mixture composed of a solid host material and a dispersed liquid. The silica minerals serve as the dispersing medium with liquid water suspended within the solid substance. Submicroscopic spheres of the silica minerals act as a three-dimensional diffraction grating, producing a spectacular array of colors. The water content is variable from 1 to 30%, with typical levels at 4–9%.[2] The water content is related to the temperature of the host rock at the time the opal formed.[3] While opal is fundamentally colorless, impurities often impart color from the yellows, oranges, and reds typical of "fire opal" derived from iron oxides to the very dark gray, blue, or black color of the rare "black opal" arising from manganese oxides and carbon.[4]

Pearls are also considered to be gemstones composed mainly of calcium carbonate ($CaCO_3$). The luster associated with pearls is a consequence of the solid emulsion nature of pearls, with water again serving as the dispersed substance.

The Chemical Details

All silica minerals have a basic three-dimensional microscopic structure composed of tetrahedral arrangements of oxygen atoms about a central silicon atom, with each oxygen atom shared with a neighboring tetrahedral group. The packing of the tetrahedra vary among the silica minerals, with the most common form of silica — *quartz* — exhibiting a relatively dense packing. Two other forms of silica — *cristobalite* and *tridymite* — have a much more open framework, easily accommodating small amounts of impurities within the three-dimensional structure. The gem opal contains cristobalite, tridymite, or mixtures of these two silica minerals. Since the structure of opal is not crystalline but *amorphous* (lacking in a periodic three-dimensional arrangement of atoms), opal is often considered a *mineraloid* (a naturally occurring "mineral" lacking a crystalline structure). Due to its varying water content, opal is also described as hydrated silicon dioxide or a hydrate of silica with a general formula of $SiO_2 \cdot nH_2O$. Using a technique known as electron microscopy, the microstructure of precious opal was first revealed in 1963.[5] The iridescent colors associated with opal arise from a semiperiodic arrangement of uniform microspheres of hydrated SiO_2 spaced at about the wavelength of visible light. These microspheres range in size from 200 to 300 nm in diameter[6] and serve as a diffraction source for refracted and reflected light.

Thermal analysis techniques reveal that water is bound in opal in more than one manner.[7] Most of the water is physically held in inclusions or microscopic pores within the opal, that is, in spaces between the microspheres. Water held in this manner can escape through complex systems of microscopic fissures or cracks, induced by temperatures even below 100°C. Some water is held within the opal via chemical bonding ("adsorption") to the surfaces of the silica microspheres and is retained to temperatures approaching 1000°C.[7] Furthermore, since the microspheres themselves are composed of much smaller silica particles, water is additionally coated on the surfaces of these minute particles. The porous nature of opal and its thermal sensitivity require special care, for dehydration may result in cracking that greatly diminishes the value of this gemstone.

KEY TERMS:	amorphous	crystalline	diffraction	hydrate
	solid emulsion			

References

[1] "The Silicate Class." Amethyst Galleries, Inc.,
http://mineral.galleries.com/minerals/silicate/class.htm

[2] http://www.eb.com:180/cgi-bin/g?DocF=macro/5004/28/94.html

[3] "The Gemstone, Opal." Amethyst Galleries, Inc.,
http://mineral.galleries.com/minerals/mineralo/opal/opal.htm

[4] "Opal." http://www.eb.com:180/cgi-bin/g?DocF=micro/439/2. html

[5] "Bibliography on Opal, Cristobalite and Related Literature." Deane K. Smith,
http://www.osha-slc.gov/SLTC/silicacrystalline/rosem/docs/ref.doc

[6] "The Mexican Opal." Carmen Ramirez, Gemmology Canada, Canadian Institute of Gemmology, http://www.cigem.ca/415.html#SCRL5

[7] "The Story of Opals." Opals International Jewellers, Inc.,
http://shop.store.yahoo.com/opals/storyofopals.html

Related Web Sites

▶ "Agates and Opals." Lecture 16b of Geology 306 , The University of Wisconsin Geology and Geophysics Department, Professor Jill Banfield,
http://www.geology.wisc.edu/~jill/Lect16b.html

- ▶ "Gems and Precious Stones." The University of Wisconsin Geology and Geophysics Department, Professor Jill Banfield, http://www.geology.wisc.edu/~jill/306.html

- ▶ "Black Opal With Brilliant Vivid Colors." Smithsonian Gem and Mineral Collection, http://www.min.uni-bremen.de/sgmcol/gems/opal.shtml

- ▶ "Pearls — More than a Gemstone?" Gemmology Canada — Special Edition, Shirley Ford-Bouchard, http://www.cigem.ca/417.html

Chapter 4 ►

Connections to Solids

4.1 Why Does Morton Salt Claim *"When It Rains It Pours"*?

Together with her slogan, "When It Rains It Pours," the Morton Salt Girl carrying her familiar umbrella during a rain shower is well recognized by today's consumers. This image was introduced by the Morton Salt Company to promote a chemical innovation in this important household product.

The Chemical Basics

In 1911 the Morton Salt Company introduced the slogan "When It Rains It Pours" with their familiar little girl with an umbrella to promote, in their words, "a packaging innovation" that allowed salt to pour easily even in humid conditions.[1] In fact, their innovation is actually an additional ingredient or "free-flowing agent"[1] that prevents salt from absorbing water and caking. Such an ingredient has the capacity to absorb many times its weight in water without itself dissolving, allowing the table salt to pour freely from its container.

The Chemical Details

What was the first free-flowing agent added to Morton Salt in 1911? Magnesium carbonate, $MgCO_3$, was the first anticaking agent used by the Morton Salt Company.[2] The extra component present in Morton Salt packages today is calcium silicate, $CaSiO_3$. This white powder has the incredible ability to absorb liquids and still remain a free-flowing powder. In general, calcium silicate absorbs 1 to 2.5 times its weight of liquids. For water, its total absorption power is estimated as 600%, that is, absorbing 600 times its weight of water.[3] Morton

International indicates that the amount of calcium silicate added to a package of table salt is "less than one-half percent" (by weight).[4] In addition to adding calcium silicate to table salt, this anti-caking agent is also included in formulations of baking powder.

KEY TERMS: anticaking agent

References

[1] Morton Salt, Morton International http://www.mortonsalt.com/

[2] General FAQs, Morton(r) Salt, Morton International
      http://www.mortonsalt.com/faqs/fprflfaqs.htm

[3] *The Merck Index*, 10th ed. (Rahway, NJ: Merck, 1983), 234.

[4] Table Salt FAQs, Morton Salt, Morton International,
      http://www.mortonsalt.com/faqs/fatsfaqs.htm

Related Web Sites

▶ Morton Salt, Morton International, http://www.mortonsalt.com

▶ The Salt Institute, http://www.saltinstitute.org

4.2   What Are "Desiccants" and Why Are They in Packaged Products?

Have you ever wondered about the content of the small packets of granules included in boxes of new electronic devices, leather products, or medications? Or have you simply heeded the warning label to dispose of the sachet immediately? The granular desiccants in these packets have particular physical properties that enhance their function as drying agents. A look at the chemical structure or chemical properties of these materials provides a better understanding of their capacity to control moisture.

The Chemical Basics

Perhaps you have just purchased a new electronic device, such as a personal computer, video camera, pager, or cell phone. The advanced circuitry in such devices can malfunction in humid environments. Moisture can reroute and even short circuit electric signals, impairing the operation of your new electronic equipment. Manufacturers recognize that moisture can adversely affect their products and therefore include small packets of "desiccant" to control the moisture levels during shipping and storage. A desiccant is a porous solid drying agent or hydrating agent that attracts moisture from the atmosphere and acts by retaining those particles of water on its surface or in the pores of the desiccant. Both synthetic materials and naturally occurring minerals function as desiccants, including dehydrated gypsum, calcinated lime, and a type of clay. In addition to electronic devices, a myriad of other products benefit from the application of desiccants, including food, pharmaceuticals, shoes and other leather articles, laboratory equipment and hygroscopic chemicals, books and rare manuscripts, museum and historical artifacts, paintings and valuable art objects, film, hearing aids, and stamps.

The Chemical Details

Five common desiccant materials are used to adsorb water vapor: montmorillonite clay ($[(Na,Ca_{0.5})_{0.33}(Al,Mg)_2Si_4O_{10}(OH)_2 \cdot nH_2O]$), silica gel, molecular sieves (synthetic zeolite), calcium sulfate ($CaSO_4$), and calcium oxide (CaO). These desiccants remove water by a variety of physical and chemical methods: *adsorption*, a process whereby a layer or layers of water molecules adhere to the surface of the desiccant; *capillary condensation*, a procedure whereby the small pores of the desiccant become filled with water; and *chemical action*, a procedure whereby the desiccant undergoes a chemical reaction with water.

Montmorillonite clay is a naturally occurring adsorbent that swells to several times its original volume when water adsorption occurs.[1] The most commonly used desiccant is silica gel ($SiO_2 \cdot H_2O$), an amorphous form of silica manufactured from sodium silicate and sulfuric acid. The porous nature of silica gel forms a vast surface area that attracts and holds water by both adsorption and capillary condensation, allowing silica gel to adsorb about 40% of its weight in water.[2] Zeolites or "molecular sieves" are rigid, hydrated crystalline aluminosilicate minerals that contain alkali and alkaline earth metals. Zeolites possess a three-dimensional crystal lattice structure that forms surface pores of uniform diameter and contain numerous regular internal cavities and channels. Water molecules are readily incorporated within the pores and cavities. The zeolite structure consists of interlocking tetrahedrons of $[SiO_4]^{4-}$ and $[AlO_4]^{5-}$. Each oxygen atom is shared by two tetrahedra. For charge balance, other metal ions are present in the cavities of the framework, typically monovalent or divalent ions such as sodium, potassium, magnesium, calcium, and barium. The drying effect of calcium sulfate (dehydrated gypsum) occurs by chemical action, as the hemihydrate of calcium sulfate is created when the anhydrous calcium sulfate reacts with water:

$$2\ CaSO_4 + H_2O \rightarrow 2\ CaSO_4 \cdot \frac{1}{2}H_2O.$$

Calcium chloride (also known as calcinated lime) acts as a drying agent as a consequence of its *deliquescent* nature — solid $CaCl_2$ readily absorbs water from the atmosphere and subsequently dissolves.

KEY TERMS: desiccant   anhydrous   hydrate   deliquescent

References

[1] The Mineral MONTMORILLONITE, Amethyst Galleries, Inc.,
     http://mineral.galleries.com/minerals/silicate/montmori/montmori.htm

[2] "Selecting the Right Desiccant." http://www.multisorb.com/faqs/selecting.html

Related Web Sites

▶ "Protection of Pharmaceuticals and Diagnostic Products through Desiccant Technology."
   Rodney L. Dobson, Multisorb Technologies Inc.,
   http://www.multisorb.com/faqs/protection.html

▶ "Frequently Asked Questions." Desiccare, Inc., http://206.190.82.22/faq.htm

▶ "The Zeolite Group." Amethyst Galleries, Inc.,
   http://galleries.com/minerals/silicate/zeolites.htm

▶ "Zeolite: The Versatile Mineral." ZeoponiX, Inc.,
   http://www.zeoponix.com/html/body_zeolites.html

▶ "Components of printing inks." Fachhochschule Stuttgart, Hochschule fur Druck and Me-
   dien http://www2.hdm-stuttgart.de/projekte/printing-inks/p_compo0.htm

Other Questions to Consider

5.18   What is the origin of the expression "in the limelight"? *See* p. 64.

8.1   What do meteorologists use to seed clouds? *See* p. 105.

12.4   What causes the pearlescent appearance of some paints? *See* p. 147.

CRYSTAL STRUCTURES

4.3   Why Is the Hope Diamond Blue?

Diamonds are the only gemstone whose colorlessness enhances their value. However, the rare, rich, natural coloring of "fancy color" diamonds commands the highest prices. The Hope Diamond possesses exceptional blue coloring and is undoubtedly the most celebrated diamond in the world. What is the origin of its intensely prized blue hue?

The Chemical Details

Throughout history civilization has treasured the rarity and beauty of fancy colored diamonds. The stunning diamond from India known as the "Hope Diamond," once a part of many royal inventories, is now the premier attraction of the Smithsonian Institution (see color Fig. 4.3.1). While the size of the diamond at 45.52 carats has certainly contributed to the public's interest in the gem, the intense blue-violet color of the stone is generally considered to be its most captivating feature. First described in the mid 1600s by the French merchant traveller Jean Baptiste Tavernier as "un beau violet" (a beautiful violet), the gem also acquired the title "Blue Diamond of the Crown" or the "Royal French Blue" when in possession of King Louis XIV of France. The blue color is attributed to trace amounts of boron in the carbon matrix of the stone. Substitution of carbon atoms by nitrogen leads to yellow diamonds, such as the famous canary yellow 128.51-carat "Tiffany" diamond.

The Chemical Details

The chemical composition and crystal structure of a mineral determine its physical and optical properties. The diamond crystalline lattice structure (Fig. 4.3.2)

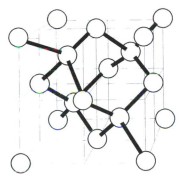

Figure 4.3.2 ▶ The diamond crystalline lattice structure composed of two interpenetrating face-centered cubic lattices.

is composed of two interpenetrating *face-centered cubic* (fcc) lattices. These lattices are characterized by equal lengths of all sides of the lattice (hence a "cubic" structure) and by atoms positioned at each corner of the cubic lattice and at the center of each face of the lattice. The two interpenetrating lattices are displaced by 1/4 of the distance of the lattice edge. Each carbon atom in one fcc lattice is tetrahedrally coordinated with four other carbon atoms in the other fcc lattice. Other materials with the diamond crystal structure include silicon and germanium. Carbon's neighbors in the periodic table, boron and nitrogen, have similar atomic sizes to carbon and can substitute for carbon in the diamond structure. Even a substitution in the parts per million range is sufficient to change the optical properties of diamond from a colorless gem to a pale blue color. Increasing substitution of boron deepens the blue hue.

> **KEY TERMS:**   **face-centered cubic lattice**

Related Web Sites

▶ "Smithsonian FAQ's: Hope Diamond." National Museum of Natural History, Smithsonian Institution, http://www.si.edu/resource/faq/nmnh/hope.htm

▶ "Introduction to Fancy Color Diamonds." Florida Jewelry Appraisers, http://www.colored-diamonds.com/page1.htm

▶ "Gems and Precious Stones: Color in Minerals." Jill Banfield, The Department of Geology and Geophysics, University of Wisconsin-Madison, http://www.geology.wisc.edu/~jill/Lect7.html

▶ "The Nature of Diamonds." American Museum of Natural History, http://www.amnh.org/exhibitions/diamonds/

▶ "Famous Fancy Diamonds: A Brief History." Gemmology Canada — Special Issue, Linda Crane, http://www.cigem.ca/423.html

Other Questions to Consider

| 3.10 | Why are opals and pearls iridescent? *See* p. 25.

| 8.1 | What do meteorologists use to seed clouds? *See* p. 105.

| 8.5 | What nineteenth-century "disease" destroyed cathedral organ pipes? *See* p. 113.

Chapter 5 ▶

Connections
to Chemical Reaction Types

OXIDATION—REDUCTION

5.1 Why Is There Abundant White Smoke from the Space Shuttle Booster Rockets on Lift-Off?

The thunderous roar of a rocket's engines and the tremendous white clouds of smoke that accompany lift-off — these sensations evoke powerful emotions in any one viewing a Space Shuttle launch. The chemistry of the propellant mixture is responsible for the billowy clouds as the Shuttle soars into the sky.

The Chemical Basics

The solid rocket boosters of the Space Shuttle are appropriately named for the solid propellant loaded within them. The ignition provided by the reaction of the solid aluminum powder and ammonium perchlorate powder generates a finely divided white powder known as *alumina*, various gases, and an extensive amount of heat. The dispersal of the white powder in the gases streaming from the boosters creates the billowy white appearance.

The Chemical Details

The propellant mixture in each solid rocket booster of the Space Shuttle contains ammonium perchlorate ("the oxidizer," 69.6% by weight), aluminum ("the fuel," 16% by weight), an iron oxide catalyst (0.4% by weight), a polymeric binder that holds the mixture together (12.04% by weight), and an epoxy curing agent (1.96% by weight)[1]. Two of the oxidation–reduction reactions that occur are

$$2 NH_4ClO_4 (s) + 2 Al (s) \rightarrow Al_2O_3 (s) + 2 HCl (g) + 2 NO (g) \quad (1)$$
$$+3 H_2O (g)$$

$$6 NH_4ClO_4 (s) + 10 Al (s) \rightarrow 5 Al_2O_3 (s) + 6 HCl (g) + 3 N_2 (g) \quad (2)$$
$$+9 H_2O (g).$$

The solid white product, alumina, Al_2O_3, is dispersed in the gaseous products, creating the billowy white exhaust plumes characteristic of Shuttle launches. Notice that, in the formation of alumina, three elements — aluminum, nitrogen, and chlorine — undergo changes in oxidation number. In both reactions (1) and (2), aluminum is oxidized from the zero oxidation state to the $+3$ oxidation state and chlorine is reduced from the $+7$ to the -1 oxidation state. The extent of oxidation of nitrogen differs in the two processes. In reaction (1) nitrogen undergoes a more extensive oxidation state change from -3 to $+2$, while in reaction (2) nitrogen experiences a more restricted oxidation from -3 to the zero oxidation state.

KEY TERMS: oxidation–reduction   oxidation state

References

[1] "Solid Rocket Boosters." NASA Space Shuttle Reference Manual, http://science.ksc.nasa.gov/shuttle/technology/sts-newsref/srb.html

Related Web Sites

▶ "The Space Shuttle: Solid Rocket Boosters." Student Space Awareness — National Web Team, University of Arizona Chapter of Students for the Exploration and Development of Space, http://seds.lpl.arizona.edu/ssa/docs/Space.Shuttle/srb.shtml

▶ "Chapter 6. Launch Systems and Launch Sites. Section 1: Principles of Rocket Propulsion." Army Space Reference Text, US Army Space Institute, Fort Leavenworth, KS, http://www.fas.org/spp/military/docops/army/ref_text/chap6im.htm

5.2 Why Do Homemade Copper Cleaners Use Vinegar?

The blue-green coating or "patina" on many bronze statues and copper artifacts often adds to the authenticity of the art. Less desirable

are the dark green and black coatings that develop on cooking utensils or jewelry items. How does understanding the chemistry of the
copper corrosion layer enable you to devise a homemade recipe for
cleaning?

The Chemical Basics

One simple yet successful technique for cleaning green, faded copper utensils,
brass items, and chrome surfaces recommends the use of vinegar: "Pour vinegar
and salt over copper surface and rub." Even black- or green-coated specimens
of native copper can be cleaned with vinegar. Alternative recipes suggest lemon
juice, ketchup, or even water in which onions have been boiled. Why do these
remedies work?

Copper is a metallic element; brass is an alloy or mixture of the metallic elements copper and zinc. The surfaces of copper and brass items tarnish with
prolonged exposure to air, particularly in moist environments with high carbon
dioxide (CO_2) or sulfur dioxide (SO_2) concentrations (see color Fig. 5.2.1). The
compounds that form on the surface, ranging in color from black to blue to dark
green, dissolve readily in acidic solutions. Vinegar contains acetic acid, ketchup
contains tomatoes rich in ascorbic acid (Vitamin C), and onions contain malic acid
and citric acid. All of these foods provide variable amounts of acid to dissolve the
tarnish on copper surfaces.

The Chemical Details

What is the identity of the corrosion layer (called *patina*) on copper metal surfaces? The oxidation of copper metal can be induced by a number of substances
present in our atmosphere, including oxygen, carbon dioxide, sulfur dioxide, and
hydrogen sulfide. Copper oxides, sulfate and carbonate salts of copper, and copper sulfides result. In seacoast locations, chloride salts may form an essential part
of the patina film. Some specific copper oxidation products include green malachite $CuCO_3 \cdot Cu(OH)_2$, blue azurite $2CuCO_3 \cdot Cu(OH)_2$, chalcocite Cu_2S, covel-
lite CuS, chalocopyrite $CuS \cdot FeS$, bornite $FeS \cdot 2 Cu_2S \cdot CuS$, tetrahedrite $4Cu_2S$
$\cdot Sb_2S_3$, brochantite ($CuSO_4 \cdot 3Cu(OH)_2$, and atacamite ($CuCl_2 \cdot 3Cu(OH)_2$.[1]
The green- and blue-colored patinas are generally composed of cupric carbon-
ates, including green malachite and blue azurite. Other green coatings are generally combinations of copper sulfate (brochantite) and copper chloride (atacamite).
Black coatings on copper are cupric oxide (copper (II) oxide or the mineral known
as tenorite, CuO), while red corrosion layers arise from cuprous oxide (copper (I)
oxide or cuprite, Cu_2O).

Recently the patina formed in the atmosphere on the roof of the Stockholm
City Hall was analyzed.[2] Several components of the patina were identified, including brochantite ($CuSO_4 \cdot 3Cu(OH)_2$), antlerite ($Cu_3(OH)_4$-SO_4), and basic
cupric carbonate ($Cu_2CO_3(OH)_6H_2O$). At the Rodin Museum in Philadelphia, active corrosion of Rodin's *The Thinker* was ascribed in 1992 to primarily brochan-

tite ($CuSO_4 \cdot 3Cu(OH)_2$), the basic copper sulfate.[3] This corrosion was attributed to an "acid rain" atmosphere as a consequence of the display of the sculpture in an outdoor urban-industrial environment for more than sixty years. During the major restoration of the Statue of Liberty in 1986, the variable blue discoloration on the copper skin of the surface was attributed to a number of minerals. A phase diagram was constructed to assign the colors on the statue's surface to particular minerals that had formed according to the varying exposure to sea spray, rainwater, and sulfur dioxide air pollution.[4] The various blue shades were assigned to brochantite ($CuSO_4 \cdot 3Cu(OH)_2$, antlerite ($Cu_3(OH)_4SO_4$), and chalcanthite (hydrated copper sulfate, $CuSO_4 \cdot 5H_2O$) and reddish coloring was ascribed to cuprous oxide (copper (I) oxide or the mineral known as tenorite). These analyses clearly demonstrate that the exposure of copper metal to the atmosphere results in complex layers of minerals containing oxides, hydroxides, sulfates, carbonates, chlorides, and sulfides.

KEY TERMS: oxidation   alloy

References

[1] "Copper Corrosion Study." F. Romero, California State University — Dominguez Hills, http://tardis.csudh.edu/fromero/chemistry/copper/index.html

[2] "Photoacoustic FT-IR Spectroscopy of Natural Copper Patina." Alex O. Salnick and Werner Faubel, Photothermal and Optoelectronic Diagnostics Laboratory, University of Toronto, Department of Mechnical Engineering, Toronto, Ontario M5S 1A4, Canada, Appl. Spec. v49 (10) (1998), http://www.s-a-s.org/journ/ASv49n10/ASv49n10_sp17.html

[3] "The Conservation of Rodin's The Thinker." Philadelphia Museum of Art, http://www.philamuseum.org/collections/conservation/projects/rodin/1.shtml

[4] "Transferring Technology from Conservation Science to Infrastructure Renewal." Richard A. Livingston, http://www.tfhrc.gov/pubrds/summer94/p94su18.htm

Related Web Sites

► "Cleaning Native Copper." Herb Sulsky, Lithosphere, May 1993, Fallbrook Gem and Mineral Society, http://fgms.home.att.net/copper.htm

► "Copper: The Red Metal." The Geology Project Homepage, Tom Lugaski, University of Nevada, Reno, http://www.unr.edu/sb204/geology/coptext.html

► "The Restoration and Conservation of Ancient Copper Coins." Doyle W. Lynch , Baylor University, http://www2.dcci.com/dlynch/digbible/restoration.shtml

► "Conservation of Nonferrous Metals." http://128.174.5.51/denix/Public/ES-Programs/Conservation/Underwater/5-CU-AG.html

► "Conservation of Cupreous Metals (Copper, Bronze, Brass)." http://nautarch.tamu.edu/class/anth605/File12.htm

▶ "The Mineral Brochantite." The Mineral and Gemstone Kingdom,
   http://mineral.galleries.com/minerals/sulfates/brochant/brochant.htm

▶ "The Mineral Antlerite." The Mineral and Gemstone Kingdom,
   http://mineral.galleries.com/minerals/sulfates/antlerit/antlerit.htm

| 5.3 | # Why Is "Fighting" Fires with Water or CO_2 Dangerous? |

blocks with a flint rod molded into the block are sold as emergency fire starters. When the magnesium is shaved into very small pieces, the shavings readily ignite with a flint spark to generate an intense (5400°F) fire. The temperatures are sufficient to ignite even damp wood. While the shavings burn rapidly and thereby disappear, what chemical principles should you understand to douse a fire involving larger amounts of magnesium?

The Chemical Basics

Magnesium metal burns vigorously in oxygen to produce a brilliant white flame. One common example of the combustion of magnesium and oxygen is a flash-bulb — the intense flash is produced by the reaction of magnesium ribbon with oxygen. However, other atmospheres support the burning of magnesium, including nitrogen, carbon dioxide, and water. Some of these reactions are even more vigorous than the reaction of magnesium and oxygen, generating heat and further exacerbating the situation.

Combustible metals, such as magnesium, titanium, zirconium, potassium, and sodium, generate what are classified as *Class D fires*. These materials burn at high temperatures and will react violently with water or other chemicals. Such materials should be handled with care. To extinguish fires involving these flammable metals, metal/sand extinguishers should be used. These extinguishers work by simply smothering the fire.

The Chemical Details

The overall balanced chemical equations for the reaction of solid magnesium with oxygen gas, nitrogen gas, liquid water, and carbon dioxide gas appear below. In each reaction, magnesium undergoes an oxidation process, i.e., an increase in oxidation number. In addition, each of these reactions is an exothermic process, releasing sizable amounts of heat at constant pressure. The heat generated by these reactions continues to fuel the combustion of magnesium, intensifying the fire:

$$2\,Mg\,(s) + O_2\,(g) \;\rightarrow\; 2\,MgO\,(s) \qquad\qquad \Delta H^\circ_{rxtn} = -1203\;kJ$$
$$3\,Mg\,(s) + N_2\,(g) \;\rightarrow\; Mg_3N_2\,(s) \qquad\qquad \Delta H^\circ_{rxtn} = -461\;kJ$$
$$Mg\,(s) + 2\,H_2O\,(l) \;\rightarrow\; Mg(OH)_2\,(aq) + H_2\,(g) \qquad \Delta H^\circ_{rxtn} = -355\;kJ$$
$$2\,Mg\,(s) + CO_2\,(g) \;\rightarrow\; 2MgO\,(s) + C\,(s) \qquad\qquad \Delta H^\circ_{rxtn} = -810\;kJ.$$

KEY TERMS: combustion   oxidation   exothermic reaction

5.4 Why Is Hydrogen Peroxide Kept in Dark Plastic Bottles?

In addition to protection from breakage, the brown plastic bottles in which hydrogen peroxide is commonly sold enhance the shelf-life of this product. Packaging plays an important role in limiting the light-induced chemical reactions of hydrogen peroxide.

The Chemical Basics

Hydrogen peroxide is a major industrial chemical essential to the organic chemicals industry and also useful in environmental treatment of polluted waters. Dilute aqueous solutions of hydrogen peroxide, H_2O_2, are commercially available for uses as a mild antiseptic (a 3% solution, i.e., 3 g of hydrogen peroxide per 100 g of solution) or bleach (a 6% solution). Hydrogen peroxide is the simplest member of the *peroxide* class of compounds, a class of chemical compounds in which two oxygen atoms are linked together by a single covalent bond. Hydrogen peroxide decomposes into water and oxygen upon heating or in the presence of trace amounts of metal ions or metal. Even traces of alkali metal ions dissolved from glass can cause this decomposition, and, for this reason, solutions of H_2O_2 are normally stored in wax-coated or plastic bottles.

The Chemical Details

The chemical structure of hydrogen peroxide in Fig. 5.4.1 depicts the oxygen−

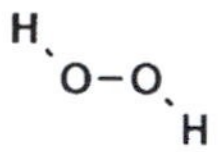

H
    O−O
        H

Figure 5.4.1 ▶ The hydrogen−oxygen−oxygen−hydrogen linkage in hydrogen peroxide.

oxygen linkage. A skew configuration exists (color Fig. 5.4.2) with a dihedral angle sensitive to the extent of hydrogen bonding in solution. In the gas phase a dihedral angle of 111.5° exists, while a value of 90.2° is measured for crystalline H_2O_2.[1]

The peroxide family of compounds is not only characterized by the oxygen–oxygen single bond but also by an oxidation state for oxygen that is intermediate between the values observed for oxygen in oxygen gas (O_2) and water (H_2O). The unusual -1 oxidation state of oxygen in hydrogen peroxide ensures that H_2O_2 can act as both a reducing agent and an oxidizing agent. When H_2O_2 acts as a reducing agent, the oxygen atoms are oxidixed from the -1 oxidation state to the 0 oxidation state in O_2:

$$H_2O_2 \text{ (l)} \rightarrow O_2 \text{ (g)} + 2 \text{ H}^+ \text{ (aq)} + 2 \text{ e}^-.$$

Alternatively, as an oxidizing agent, H_2O_2 yields H_2O (or OH^- in basic solution) as the oxygen atoms undergo reduction to the -2 oxidation state:

$$H_2O_2 \text{ (l)} + 2 \text{ H}^+ \text{ (aq)} + 2 \text{ e}^- \rightarrow 2 \text{ H}_2O \text{ (l)}.$$

Under certain conditions H_2O_2 can simultaneously undergo both oxidation and reduction — i.e., *disproportionation* — to yield water and oxygen:

$$2 \text{ H}_2O_2 \text{ (l)} \rightarrow 2 \text{ H}_2O \text{ (l)} + O_2 \text{ (g)}.$$

While this *decomposition reaction* is slow at room temperature, the disproportionation process can be accelerated by heat, certain catalysts (e.g., metal ions), and light. Traces of alkali metal ions are normally present in aqueous solutions stored in glass bottles, so plastic or wax-coated glass containers reduce the concentration of metal ion catalysts. Brown-colored containers limit the wavelengths of light that can be absorbed by the solution, restricting the initiation of the disproportionation reaction.

KEY TERMS:	disproportionation	oxidizing agent	reduction	peroxide
	oxidation state	reducing agent	oxidation	

References

[1] N. N. Greenwood and A. Earnshaw, "Hydrogen Peroxide," in *Chemistry of the Elements*, 2nd ed. (Oxford: Pergamon Press, 1997), 633–638.

Related Web Sites

▶ "Introduction to Hydrogen Peroxide." H2O2.com, Southland Environmental and FMC Corporation, http://www.h2o2.com/intro/overview.html

5.5 Why Does a Flashbulb Develop a White Coating after a Flash?

A... brilliant light... can be obtained by burning ... magnesium in oxygen. A piece of magnesium wire held by one end in the hand, may be lighted at the other extremity by holding it to a candle... It then burns away of its own accord evolving a light insupportably brilliant to the unprotected eye... .

—William Crookes, editor of
Photographic News, Oct. 1859[1]

The Chemical Basics

Flashbulbs and flashlamps are devised to produce intensely brilliant but brief emissions of light for a variety of applications. Photographic flashbulbs contain a mesh of fine magnesium wire in an oxygen-rich atmosphere. When the flashbulb is activated, an electric current passes through the wire, heating the metal, triggering a reaction with oxygen to produce a white, powdery solid known as magnesium oxide that coats the interior of the bulb (see color Fig. 5.5.1).

The Chemical Details

The first portrait using magnesium was taken by Alfred Brothers of Manchester on 22 February, 1864.[1] However, a limited understanding of the chemistry of magnesium and its prohibitive cost inhibited the development of this application. Advances by Robert Bunsen (the discoverer of the element cesium and for whom the Bunsen burner is named) led to more economic production of pure magnesium by electrolytic means.[2] Bunsen also advocated the use of magnesium for illumination purposes. In the late 1880s the ignition of magnesium powder with an oxidizing agent such as potassium chlorate was discovered and led to the introduction of flash powder. Professional photographers then produced brilliant flashes of light (as well as acrid smoke and ash) by firing finely powdered magnesium with a percussion cap.

The development of photographic flashbulbs occurred in the 1920s. These devices generally consist of a transparent glass bulb filled with a mesh of aluminum, magnesium, or zirconium metal in a volume of oxygen gas. When these metals burn in air, i.e., undergo combustion with oxygen, they produce an intense white light — the flash we observe in a flash bulb. However, the combustion of these metals with oxygen is a slow process at room temperature. Heating of a thin metal filament by an electric current produces the temperatures needed for a rapid reaction. For example, zirconium combines chemically with oxygen at 800°C to produce the white solid zirconium oxide, ZrO_2. Magnesium oxide, MnO, and aluminum oxide, Al_2O_3, are also white solids. The balanced chemical equations for these oxidation or combustion reactions are summarized below:

$$2\,Mg\,(s) + O_2\,(g) \;\rightarrow\; 2\,MgO\,(s)$$
$$Zr\,(s) + O_2\,(g) \;\rightarrow\; ZrO_2\,(s)$$
$$4\,Al\,(s) + 3\,O_2\,(g) \;\rightarrow\; 2\,Al_2O_3\,(s).$$

KEY TERMS: combustion   oxidation

References

[1] "A History of Photography: Lighting." Robert Leggat, 1996,
    http://www.rleggat.com/photohistory/history/lighting.htm

[2] "Robert Wilhem Bunsen (1811–1899)."
    http://www.woodrow.org/teachers/chemistry/institutes/1992/Bunsen.html

Related Web Sites

▶ "Picture This… History of Photography." http://www.picturethis.net/history.html

5.6   Why Did Dorothy Have to Oil the Tin Woodman in *The Wizard of Oz*?

"What can I do for you?" she [Dorothy] inquired softly, for she was moved by the sad voice in which the man spoke.

"Get an oil-can and oil my joints," he answered. "They are rusted so badly that I cannot move them at all; if I am well oiled I shall soon be all right again."

—L. Frank Baum, *The Wonderful Wizard of Oz*

The Chemical Basics

Had the Tin Woodman been constructed of pure tin he would not have rusted. Typically, the term "rust" is reserved for the product of the oxidation of the metal iron or its alloys, often due to atmospheric conditions. The Tin Woodman most likely was constructed of the same type of material used for "tin cans"— *tinplate* — a thin sheet of iron or steel (an iron alloy) coated with tin. The iron component of the Woodman's framework oxidized in air to produce the product iron oxide or rust.

The Chemical Details

The oxidation of iron by oxygen is a spontaneous process in acidic, neutral, or basic environments. Iron serves as the anode and reducing agent as it is oxidized to Fe^{2+}:

$$Fe\ (s) \rightarrow Fe^{2+} + 2\ e^-.$$

Oxygen acts as the *oxidizing agent*. The products of the reduction of O_2 depend on the acidity of the environment. In acidic solution, the reduction of O_2 generates water:

$$O_2 + 4\ H^+ + 4\ e^- \rightarrow 2\ H_2O.$$

In alkaline or neutral solution, hydroxide ions are the oxidation product:

$$O_2 + 2\ H_2O + 4\ e^- \rightarrow 4\ OH^-.$$

Thus, the overall reaction in acidic solution is

$$2\ Fe\ (s) + O_2\ (g) + 4\ H^+ \rightarrow 2\ Fe^{2+}\ (aq) + 2\ H_2O\ (l).$$

In a neutral or alkaline solution, the oxidation of iron is represented by

$$2\ Fe\ (s) + O_2\ (g) + 2\ H_2O \rightarrow 2\ Fe^{2+}\ (aq) + 4\ OH^-\ (aq).$$

Thus, the primary reactions at the surface of iron are the loss of the metal due to oxidation to the divalent ion and the reduction of O_2 gas to either water or hydroxide ion. The formation of rust is actually a secondary oxidation reaction of the Fe^{2+} ions to Fe^{3+} with additional O_2, forming insoluble Fe_2O_3:

$$4\ Fe^{2+} + O_2\ (g) + 4\ H_2O \rightarrow 2\ Fe_2O_3 + 8\ H^+$$

Tin plating is a common procedure to protect iron and its alloys from rusting. Tin is less easily oxidized than iron, as revealed by the relative magnitudes of the *standard reduction potentials* of the two metals:

$$Sn^{2+} + 2\ e^- \rightarrow Sn\ (s) \qquad \xi^\circ = -0.14\ V$$
$$Fe^{2+} + 2\ e^- \rightarrow Fe\ (s) \qquad \xi^\circ = -0.44\ V.$$

As long as the tin coating remains intact, the iron is protected from oxidation. A scratch of the tin surface to expose the iron metal leads to the oxidation of iron in preference to the tin.

KEY TERMS: **oxidation   oxidizing agent   anode   reducing agent**

What Puts the "Blue" in "Blue Jeans"?

"In recognition of his services in the advancement of organic chemistry and the chemical industry, through his work on organic dyes and hydroaromatic compounds."

—Royal Swedish Academy of Sciences,
1905 Nobel Prize in Chemistry to
Johann Friedrich Wilhelm Adolf von Baeyer

While we think of "blue jeans" as the quintessential American item of clothing, it was the contributions of the German chemist Johann Friedrich Wilhelm Adolf von Baeyer that put the "blue" in blue jeans and enabled this image of American culture to flourish.

The Chemical Basics

The dye indigo is responsible for the recognizable blue color of blue jeans. This dye is both a natural dye as well as a synthetic one. Indigo plants, any shrub or herb of the genus *Indigofera*, were once cultivated for the purpose of extracting the indigo pigment. However, the research chemist Johann Friedrich Wilhelm Adolf von Baeyer synthesized indigo in 1880 and formulated its structure in 1883, an accomplishment for which he won the Nobel Prize in Chemistry in 1905. The industrial process to synthesize indigo was developed by the German firm Badische Anilin- & Soda-Fabrik, now known as BASF, placing indigo on the market in 1897. The current major producers of indigo dyestuff are Ciba Geigy, ICI, BASF, and Mitsui Touatsu in Japan. As a synthetic fiber dye, indigo possesses several characteristics that confer the well-known features of blue jeans. The dye is quite colorfast to water and light, although it continually fades in intensity over time. Furthermore, the dye's structure prevents its complete penetration into fibers, yielding the irregular and individual appearance generally valued in jeans. You've probably noticed the white specks of fiber when you look closely at denim material.

The Chemical Details

Indigo (or 2-(1,3-dihydro-3-oxo-2H-indol-2-ylidene)-1,2-dihydro-3H-indol-3-one) (Fig. 5.7.1) is an example of a water-insoluble dye, contributing to its resistance

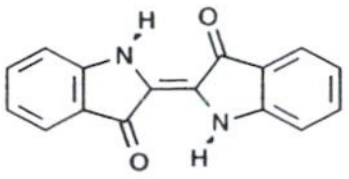

Figure 5.7.1 ▶   The molecular structure of the water-insoluble dye indigo (2-(1,3-dihydro-3-oxo-2H-indol-2-ylidene)-1,2-dihydro-3H-indol-3-one).

to fading in water and light. However, to dye denim, a cotton fabric, favorable interactions of the dye and fiber are needed. Cotton is composed of cellulose, a linear polymer with hydroxyl substituents that favor interaction with dyes via hydrogen bonding. To promote interaction with cotton, indigo is applied to textile fibers in its soluble reduced form — yellow leucoindigo or [2,2′-biindole]-3,3′-diol (Fig. 5.7.2). The added hydroxyl (–OH) groups in leucoindigo promote wa-

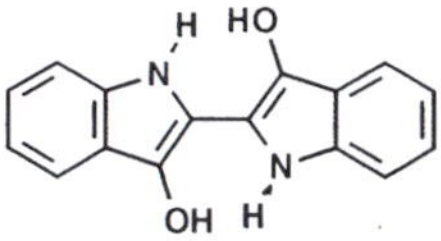

Figure 5.7.2 ▶ The soluble reduced form of the dye indigo—yellow leucoindigo or [2,2′-biindole]-3,3′-diol.

ter solubility. Sodium hydroxide (NaOH) and sodium hydrosulfite ($Na_2S_2O_4$) are capable of reducing indigo to its colorless form. "Vat dye" is the name given to the class of colored but water-insoluble dyes that must be applied in their reduced state. After dyeing the cotton fiber, the blue color is produced by reoxidizing the indigo. Exposure to air, actually atmospheric oxygen, is sufficient to oxidize the dye to its blue form. Alternatively, chromic acid (potassium dichromate and sulfuric acid) may be used as the oxidizing agent. When laundering jeans, a thimble full of vinegar added to the wash is recommended to keep the jeans dark. Vinegar (or acetic acid) acts as an acid to counterbalance or neutralize the alkaline nature of the detergent in the wash.

KEY TERMS: oxidation   reduction   hydrogen bonding

Related Web Sites

▶ "Indigo Ring Dyeing: Chemistry of Indigo Dyeing." Virkler Company, Charlotte, NC, http://www.virkler.com/expertise/html/chemistry.html

5.8 How Is Lime Used to Mitigate the Acid Rain Problem?

Buildings and statues constructed of marble are sensitive to the destructive action of acid rain. How can the same reactions that destroy marble be used to help reduce the harmful effects of acid rain pollution?

The Chemical Basics

Acid rain is caused primarily by sulfur dioxide emissions from burning fossil fuels such as coal, oil, and natural gas. Sulfur is an impurity in these fuels; for example, coal typically contains 2–3% by weight sulfur.[1] Other sources of sulfur include the industrial smelting of metal sulfide ores to produce the elemental metal and, in some parts of the world, volcanic eruptions. When fossils fuels are burned, sulfur is oxidized to sulfur dioxide (SO_2) and trace amounts of sulfur trioxide (SO_3).[2] The release of sulfur dioxide and sulfur trioxide emissions to the atmosphere is the major source of acid rain. These gases combine with oxygen and water vapor to form a fine mist of sulfuric acid that settles on land, on vegetation, and in the ocean.

One way to control gaseous pollutants like SO_2 and SO_3 is to remove the gases from fuel exhaust systems by absorption into a liquid solution or by adsorption onto a solid material. Absorption involves dissolving the gas in a liquid while adsorption is a surface phenomenon. In each case, a subsequent chemical reaction can occur to further trap the pollutant. Lime and limestone are two solid materials that effectively attract sulfur dioxide gas to their surfaces. The ensuing chemical reaction converts the gaseous pollutant to a solid nontoxic substance that can be collected and disposed or used in another industry.

The Chemical Details

When coal, oil, and gas are burned for energy in power plants and in industries, sulfur dioxide is produced. Sulfur dioxide combines with water in the atmosphere to produce sulfurous acid (H_2SO_3). Subsequent oxidation in the presence of oxygen (air) yields sulfuric acid (H_2SO_4).

$$SO_2 + H_2O \rightarrow H_2SO_3$$
$$H_2SO_3 + \tfrac{1}{2}O_2 \rightarrow H_2SO_4.$$

While the use of low-sulfur fuels is one mechanism to reduce sulfur dioxide emission, alternatively most approaches focus on "scrubbing" or ridding the emissions in smoke stacks of sulfur dioxide gas. A number of different types of "scrubbers," i.e., sulfur dioxide removal systems, are available for industry. One system sprays the flue gas into a liquid solution of sodium hydroxide. The hydroxide combines with SO_2 and O_2 to form the corresponding sulfate which can be removed from the aqueous solution:

$$2 \text{ NaOH (aq)} + SO_2 \text{ (g)} + \frac{1}{2}O_2 \text{ (g)} \rightarrow Na_2SO_4 \text{ (s)} + H_2O \text{ (l)}$$

Sodium sulfite is also used to absorb sulfur dioxide, producing a bisulfite that is converted back to sodium sulfite by the addition of a calcium precipitating agent.

Dry sorbents are also used to remove sulfur dioxide. Lime (calcium oxide) and slaked lime (calcium hydroxide) combine with sulfur dioxide to form calcium sulfite and water:

$$CaO\ (s) + SO_2\ (g)\ \rightarrow\ CaSO_3\ (s)$$
$$Ca(OH)_2\ (s) + SO_2\ (g)\ \rightarrow\ CaSO_3\ (s) + H_2O\ (l).$$

Limestone (calcium carbonate) is also used as a dry sorbent, forming calcium sulfite and carbon dioxide gas:

$$CaCO_3\ (s) + SO_2\ (g) \rightarrow CaSO_3 + CO_2.$$

Scale formation in the scrubber can lead to sodium carbonate as an additional dry sorbent in the scrubber. Alternatively, limestone is also introduced into combustion chambers to treat sulfur dioxide emissions. Decomposition of $CaCO_3$ into CaO and CO_2 occurs in the combustion chamber, and the resulting CaO combines with SO_2 to produce calcium sulfite. Notice that this process produced another potentially environmentally harmful pollutant (CO_2) as it gets rid of a definite environmentally harmful pollutant (SO_2).

New adsorbent systems use zeolites to remove the gaseous SO_2.[3] Zeolites are naturally occurring compounds that act as molecular sieves to trap SO_2. Typically microporous crystalline aluminosilicate structures, zeolites are inert materials that trap small molecules in their internal cavities. Adsorption via both physical forces (physisorption) and chemical bonding (chemisorption) is possible, depending upon the properties of the guest molecule. How is the SO_2 released or treated? By heating the spent zeolites (i.e., the zeolites with adsorbed SO_2), the SO_2 is driven off, regenerating the zeolites. The concentrated SO_2 gas can be captured and used to form byproducts, including liquefied SO_2, elemental sulfur, and ammonium sulfate, which is used as a fertilizer.

KEY TERMS: oxidation   combination reaction   zeolites   adsorption

References

[1] Geography 210: Introduction to Environmental Issues, 7.4 Air Pollution, Dr. Michael Pidwirny, Department of Geography, Okanagan University College http://www.geog.ouc.bc.ca/conted/onlinecourses/geog_210/210_7_4.html

[2] "Formation Mechanisms." Sulfur Oxides, Module 6: Air Pollutants and Control Techniques, OL 2000: An Online Training Resource: Basic Concepts in Environmental Science, North Carolina State University http://www.epin.ncsu.edu/apti/ol_2000/module6/sulfur/formation/formfram1.htm

[3] "Penn State University Researchers Discover Inexpensive Way to Clean Up Sulfur Dioxide from Coal Plants." U.S. Department of Energy, National Energy Technology Laboratory, http://www.netl.doe.gov/newsroom/media_rel/mr_pennst.html

Related Web Sites

▶ Chemical of the Week: Lime — Calcium Oxide CaO, University of Wisconsin, http://scifun.chem.wisc.edu/chemweek/lime/lime.html

▶ "Control Techniques." Sulfur Oxides, Module 6: Air Pollutants and Control Techniques, OL 2000: An Online Training Resource: Basic Concepts in Environmental Science, North Carolina State University, http://www.epin.ncsu.edu/apti/ol_2000/module6/sulfur/control/control.htm

▶ "Environmental Effects of Acid Rain." Acid Rain Program, United States Environmental Protection Agency, http://www.epa.gov/airmarkets/acidrain/index.html

▶ "Sulfur Dioxide Control." Andersen 2000 Inc., http://www.crownandersen.com/Sulfur.html

---

5.9 Why Are Bombardier Beetles Known as "Fire-Breathing Dragons"?

What chemistry is used by bombardier beetles to repel predators?

The Chemical Basics

As a defense mechanism, bombardier beetles are able to eject a hot spray to ward off their predators. This secretion provides the beetle extra time to fend off predators, for the wings of bombardier beetles take an unusually long time to unfold and limit a winged escape. The hot solution results when a chemical reaction takes place within glands in the abdomen. The bombardier beetle is able to concentrate and store hydrogen peroxide and hydroquinones in a single reservoir within the glands. When disturbed by a predator, muscular contractions force these liquids into a second chamber where certain enzymes are stored. The enzymes catalyze the decomposition of hydrogen peroxide into water and oxygen gas and convert the hydroquinones into substances known simply as quinones. These reactions generate heat, which raises the temperature of the solution ejected from the beetle's abdomen. The hot solution has been observed to be released in bursts[1] with tremendous accuracy for its target.

The Chemical Details

The decomposition of hydrogen peroxide into water and oxygen gas is a slow oxidation–reduction reaction. Hydrogen peroxide serves as both oxidizing agent and reducing agent. Under such conditions, the reaction is classified as a disproportionation reaction. Enzymes known as *catalases* can speed the rate of the disproportionation reaction. One of the products of the reaction, oxygen, can serve as an oxidizing agent to convert hydroquinones into quinones. The overall conversion of hydroquinones and hydrogen peroxide to quinones and water is an exothermic process, releasing heat. The heat is absorbed by the liquid's products (i.e., the aqueous solution of quinones), raising the temperature of the product solution. Temperatures typically on the order of 100°C are reached:

$$2 H_2O_2 \text{ (aq)} \rightarrow O_2 \text{ (g)} + 2 H_2O \text{ (l)}$$
$$2 C_6H_4(OH)_2 \text{ (aq)} + O_2 \text{ (aq)} \rightarrow 2 C_6H_4O_2 \text{ (aq)} + 2 H_2O \text{ (l)}$$
$$C_6H_4(OH)_2 \text{ (aq)} + H_2O_2 \text{ (aq)} \rightarrow C_6H_4O_2 \text{ (aq)} + 2 H_2O \text{ (l)}.$$

  hydroquinone   quinone

The enthalphy change for this reaction at 25°C can be estimated from the algebraic summation of enthalphy changes for reactions that combine to give the same overall reaction (an application of Hess' Law):

$C_6H_4(OH)_2$ (aq)	$\rightarrow C_6H_4O_2$ (aq) + H_2 (g)	$\Delta H° = +177.4 \text{ kJ}$
H_2O_2 (aq)	$\rightarrow H_2$ (g) + O_2 (g)	$\Delta H° = +191.2 \text{ kJ}$
$2\{H_2$ (g) + $\frac{1}{2}O_2$ (g)	$\rightarrow H_2O$ (g)$\}$	$\Delta H° = 2\{-241.8 \text{ kJ}\}$
H_2O (g)	$\rightarrow H_2O$ (l)	$\Delta H° = -43.8 \text{ kJ}$

$C_6H_4(OH)_2$ (aq) $\rightarrow C_6H_4O_2$ (aq) + H_2O (l) $\Delta H_{overall} = \text{sum}$
  $+H_2O_2$ (aq)

$$\Delta H_{overall} = +177.4 \text{ kJ} + 191.2 \text{ kJ} + 2\{-241.8 \text{ kJ}\} - 43.8 \text{ kJ}$$
$$\Delta H_{overall} = -159 \text{ kJ}$$

Where does the hydrogen peroxide originate? Many metabolic processes produce hydrogen peroxide as a byproduct. In particular, the degradation of amino acids and fatty acids leads to hydrogen peroxide formation.

KEY TERMS:	oxidation-reduction	disproportionation
	Hess' Law	exothermic reaction

References

[1] Jeffrey Dean, *et al.*, "Defensive Spray of the Bombardier Beetle: A Biological Pulse Jet." *Science* **248** (1990): 1219.

Related Web Sites

► "Chemical Warfare." Beth Saulnier, *Cornell Magazine On/Line* **103** (July/August 2000), http://cornell-magazine.cornell.edu/Archive/July2000/JulyBugs.html

► "Chemical Secretions of the Suborder Adephaga (Coleoptera)." Kelly B. Miller, Colorado State University, Fort Collins, CO 80523, http://www.colostate.edu/Depts/Entomology/courses/en570/papers_1996/miller.html

▶ "The Bombardier Beetle." Eric Benson,
http://www.wam.umd.edu/~ebenson/brachinus2.html

▶ "Bull's-eye Beetle." BBC Online Network, August 17, 1999,
http://news.bbc.co.uk/hi/english/sci/tech/newsid_422000/422599.stm

▶ "Secondary Metabolites (All's Fair in Love & War)." Biological Chemistry — Lecture 5,
Wayne Best, University of Western Australia,
http://www.cygnus.uwa.edu.au/~wmbest/biolchem/lecture-05.pdf

PRECIPITATION & DISSOLUTION

5.10   Why Do Seashells Vary in Color?

The striking beauty of seashells — their diverse colors, intricate patterns, and varied shapes — captures the interest of collectors all over the world. What role does chemistry play in the color of a seashell?

The Chemical Basics

Shell collecting is a favorite pastime of many beachgoers. We are intrigued by the fascinating shapes and variety of colors of the seashells that we see around the world. But what purpose does the coloration of shells serve the inhabiting animal? Colors in seashells often serve the purpose of camouflage, provide a means of communication between species, aid in temperature regulation for intertidal species, and even serve a structural function to strengthen the shell.

The basic constituent of seashells is calcium carbonate, an insoluble compound formed from calcium ions secreted from the cells of the shellfish and carbonate ions present in seawater. But calcium carbonate is a white solid. The colors of seashells often arise from impurities and metabolic waste products captured in the solid shell as it is formed. Coloration is dictated by both diet and water habitat. For example, some cowries that live and feed on soft corals take on the hue of the coral species. Yellow and red colors often arise from carotenoid pigments such as β-carotene. Light refraction often generates the iridescent "mother-of-pearl" hues.

The Chemical Details

The formation of a calcium carbonate shell is an example of a *precipitation reaction*:

$$Ca^{2+} (aq) + CO_3^{2-} (aq) \rightarrow CaCO_3 (s).$$

The insolubility of calcium carbonate is clearly evident from the value of the solubility product, K_{sp}, in water at $25°C$: $K_{sp} = 8.7 \times 10^{-9}$. The carbonate ions are produced in seawater by the dissociation of carbonic acid that forms from the reaction of dissolved carbon dioxide gas and water:

$$H_2O \text{ (aq)} + CO_2 \text{ (g)} \rightarrow H_2CO_3 \text{ (aq)}$$

$$H_2CO_3 \text{ (aq)} \rightarrow H^+ \text{ (aq)} + HCO_3^- \text{ (aq)}$$

$$HCO_3^- \text{ (aq)} \rightarrow H^+ \text{ (aq)} + CO_3^{2-} \text{ (aq)}.$$

Some of the pigments that may be found in seashells include melanin (brown and black hues), carotenoids (yellow and orange coloration), and pterodines (red shades). Many shellfish are capable of short-term coloration changes using pigment-containing cells known as *chromatophores* located in the deeper layers of the animal's skin. For example, the black pigment sepiomelanin (also known as *sepia* and the basis for sepia writing ink) isolated from the ink sac of the cuttlefish *Sepia officinalis* enables this mollusk to blend into its background.

KEY TERMS: precipitation   pigments

Related Web Sites

▶ "Why Do Shells Have Their Colors?" Gary Rosenberg, Academy of Natural Sciences, Philadelphia, http://coa.acnatsci.org/conchnet/rose0397.html

▶ "Use of Color and Light by Fish." Paul Maslin, Chico State University, http://www.csuchico.edu/~pmaslin/ichthy/Color.html

---

5.11   Why Is Vinegar Recommended for Cleaning Automatic Coffee Makers and Steam Irons?

Vinegar has been called Mother Nature's Liquid Gold. At the very least, vinegar is recognized as a "natural" and effective household cleaner. Why is vinegar effective for cleaning automatic coffee makers and steam irons?

The Chemical Basics

Vinegar is recommended for cleaning a variety of appliances and other items that may be stained by hard water deposits. Automatic coffee makers, steam irons, dishwashers, teapots, faucet heads, and shower heads — over time, all accumulate calcium deposits from hard water. Groundwater, that is, water that travels through soil and rocks, accumulates dissolved calcium ions as a consequence of the natural weathering of minerals that contain calcium such as limestone and calcite, shells, and coral. At the same time, carbon dioxide in the air dissolves in water to form carbonate ions that combine with calcium ions to form a white solid, calcium carbonate. Vinegar removes the white deposits of calcium carbonate by dissolving the solid.

The Chemical Details

Vinegar, or acetic acid, combines with calcium carbonate to dissolve the precipitate, form free calcium ions and water, and liberate carbon dioxide gas:

$$2\ CH_3COOH\ (aq) + CaCO_3\ (s)\ \rightarrow\ Ca^{2+}\ (aq) + 2\ CH_3COO^-\ (aq)$$
$$+CO_2\ (g) + H_2O\ (l).$$

The dissolution of calcium carbonate by vinegar can also be observed when a hard-boiled egg is placed in a vinegar solution. Overnight the egg shell, composed of calcium carbonate, will dissolve leaving only an outer membrane. Chalk, also composed of calcium carbonate, will also dissolve in a vinegar (or acid) solution. In preparing hard-boiled eggs to be treated with colored dyes, as in dyeing "Easter Eggs," it is often recommended to add a small amount of vinegar to the aqueous dye solution. This practice helps the dye adhere to the egg shell by creating a "fresh" surface on the egg exterior through partial dissolution of the old surface.

> **KEY TERMS:**   dissolution   precipitate

5.12   Why Does Soap Scum Form? Why Are Phosphates in Detergents?

Water softener manufacturers remind us of the many undesirable effects produced by hard water. Besides enhancing the likelihood of harmful scale deposits in plumbing, water heaters, and dishwashers, hard water also has been associated with bathtub scum, deposits on laundry, scale on glasses and dishes, scratchy skin, and unmanageable hair. A simple chemical process explains the origin of hard water. Additional chemical reactions provide an explanation for the scum and deposits that readily form when soap and hard water combine.

The Chemical Basics

More than 60% of Earth's water is groundwater, that is, water that travels through soil and rocks. Natural weathering of minerals such as limestone and calcite, shells, and coral is common as a consequence of the natural acidity of rainfall that results from dissolved atmospheric carbon dioxide. When this weathering process releases to the water supply positively charged metal ions that are constituents of minerals, the ions can combine with negatively charged ions present in soaps and detergents to form a waxy scum that does not dissolve in water.

The Chemical Details

Soap scum is an insoluble precipitate that forms between the cations of minerals typically present in hard water and the anions of soaps and detergents. Divalent cations of calcium (Ca^{2+}) and magnesium (Mg^{2+}) from calcium carbonate and magnesium carbonate minerals are the primary components of hard water. Divalent cations of iron (Fe^{2+}), manganese (Mn^{2+}), and strontium (Sr^{2+}) are also often present. An example of the dissolution (dissolving) process that releases calcium ions from calcium-containing minerals in contact with water with high acid levels is

$$CaCO_3 \text{ (s)} \quad + 2\, H^+ \text{ (aq)} \rightarrow Ca^{2+} \text{ (aq)} + H_2CO_3 \text{ (aq)}.$$
$$\text{(calcite, limestone, shells, coral)}$$

Soaps are composed of sodium salts of various fatty acids. These acids include those with the general structure $CH_3–(CH_2)_n–COOH$ where $n = 6$ (caprylic acid), 8 (capric acid), 10 (lauric acid), 12 (myristic acid), 14 (palmitic acid), and 16 (stearic acid). Oleic acid ($CH_3–(CH_2)_7–CH=CH–(CH_2)_7–COOH$) and linoleic acid ($CH_3–(CH_2)_4–CH=CH–CH_2–CH=CH–(CH_2)_7–COOH$) are also common soap ingredients. These sodium salts readily dissolve in water, but other metal ions such as Ca^{2+} and Mg^{2+} form precipitates with the fatty acid anions. For example, the dissolution of the sodium salt of lauric acid and the subsequent formation of a precipitate of the lauric acid anion with calcium ion is given by

$$CH_3–(CH_2)_{10}–COO^-Na^+ \text{ (s)} \xrightarrow[H_2O]{} CH_3–(CH_2)_{10}–COO^- \text{ (aq)} + Na^+ \text{ (aq)}$$

$$2\, CH_3–(CH_2)_{10}–COO^- \text{ (aq)} + Ca^{2+} \text{ (aq)} \rightarrow$$
$$(CH_3–(CH_2)_{10}–COO^-)_2Ca^{2+} \text{ (s)}.$$

A variety of methods can be used to soften water. A cation exchange process is common in water softeners or conditioners. In these systems, hard water containing calcium and magnesium ions is passed through resin beads composed of styrene and divinylbenzene[1] and on whose surfaces are bound potassium and sodium ions. As the hard water passes through the resin beads, calcium and magnesium ions are retained on the resin surface while sodium and potassium ions are released to the water. For charge balance, two sodium or potassium ions are released per divalent ion retained. As mentioned, the sodium and potassium cations do not form insoluble precipitates with the fatty acid anions in soap. To regenerate the beads, a strong NaCl or KCl salt water solution is flushed through the softener. Calcium and magnesium ions are discharged as waste, and sodium and/or potassium ions are returned to the surface of the beads.

An alternative method of water softening involves adding chemical compounds known as *chelators* or *sequestrants*. Many household and institutional cleaners as well as personal care products contain glycolic acid ($HOCH_2COOH$), also

known as an alpha hydroxycarboxylic acid, to complex with metal cations to reduce the hardness of water.[2] The chelates formed from glycolic acid and cations
such as calcium, magnesium, manganese, iron, and copper are water-soluble.
Both the hydroxyl and carboxylic acid groups are used to form five-member ring
complexes (chelates) with polyvalent metals. The formation of these chelated
complexes frees the detergent from precipitating with the hard water metal ions,
thereby enhancing the detergent's ability to clean. Compounds classified as *polyphosphates* also serve as sequestrants in detergents. Derivatives of phosphoric
acid (H_3PO_4) (Fig. 5.12.1) and pyrophosphoric acid ($H_4P_2O_7$) (Fig. 5.12.2) are

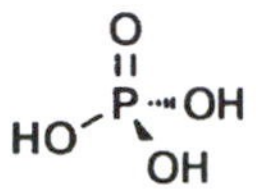

Figure 5.12.1 ▶ The molecular structure of phosphoric acid (H_3PO_4).

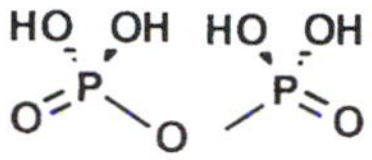

Figure 5.12.2 ▶ The molecular structure of pyrophosphoric acid ($H_4P_2O_7$).

common examples of sequestrants, specifically sodium acid pyrophosphate (disodium dihydrogen pyrophosphate, $Na_2H_2P_2O_7$), potassium (or sodium) acid
phosphate (potassium or sodium dihydrogen phosphate, KH_2PO_4 or NaH_2PO_4),
sodium hexametaphosphate (sodium polyphosphate, $(NaPO_3)_n \cdot Na_2O$), and tetra-
sodium pyrophosphate ($Na_4P_2O_7$).[3] A number of commercial products contain
such sequestering agents, including Calgon, White Rain, and Spring Rain.

Unwanted calcium and magnesium ions can also be precipitated from water by
adding washing soda ($Na_2CO_3 \cdot 10\ H_2O$), borax ($Na_2B_4O_7 \cdot 10\ H_2O$), or sodium
silicate. Many commercial products contain these precipitating agents, including
Arm and Hammer Washing Soda, Borateem, and Raindrops.

KEY TERMS: **dissolution** **precipitation** **chelator** **sequestrant**

References

[1] "What Makes Water Hard & How Hard Water Can Be Improved." Water Quality Association, http://www.bigbrandwaterfilters.com/water_treatment_info/soft_water_wqa.html

[2] "DuPont Specialty Chemicals: Glycolic Acid Applications."
http://www.dupont.com/glycolicacid/applications/

[3] "Water Treatment." Rhodia Eco Services,
        http://www.rp.rpna.com/rhodia/searchresultsmarket.icl

Related Web Sites

▶ "Soaps and Detergents: Chemistry." Soap and Detergent Association,
    http://www.sdahq.org/cleaning/chemistry//

▶ "Packaged Water Softeners." Rose Marie Tondl,
    http://www.ianr.unl.edu/pubs/water/nf94.htm

▶ "Water Softening." Morton Salt, Morton International,
    http://www.mortonsalt.com/soft/prflsoft.htm

▶ "Chemical of the Week: Phosphoric Acid." University of Wisconsin,
    http://scifun.chem.wisc.edu/chemweek/h3po4/H3PO4.html

5.13   Why Do Old Paintings Discolor?

Art conservation, particularly painting restoration, is an important en-
deavor to preserve our cultural heritage and maintain the aesthetic
value of an artistic piece. Chemical reactions occurring on a microscopic level are the origin of the macroscopic changes that we observe as ageing.

The Chemical Basics

Artists incorporate a variety of pigments in their paints to provide the color necessary to express their creative intentions. To maintain the artist's colors, museums
and art collectors know the importance of such environmental factors as temperature, relative humidity, light intensity, and air quality. Nevertheless, the nature of
the work of art itself may influence the ageing properties of the paint. The origin
and purity of the pigments used, the combination of pigments selected, the pigment volume applied, and the type of binding medium utilized (e.g., drying oils
such as linseed) are factors that influence the painting's permanence.[1] Throughout history, many painters have been well aware of the ageing properties of their
paints and took precautions to prevent these changes.[2] Nevertheless, unantici-
pated chemical reactions can occur between the materials in the painting and as
a consequence of environmental exposure. In response to the new material environment that forms as the painting ages, additional changes in the physical and
chemical properties of the pigments can subsequently occur to discolor the work
further.

The Chemical Details

A study of the chemical aspects of the ageing of works of art is a vast undertaking.
Many of the investigations focus on the artist's selection of pigments. In partic-

ular, a variety of lead-containing pigments have been used by artists throughout the centuries. Lead-containing white pigments include *carbonate white lead* used by the ancient Egyptians, Greeks, and Romans with a variable composition generally designated as $2 \, PbCO_3 \cdot Pb(OH)_2$. *Basic lead sulfate* is a white pigment also of variable composition such as $PbSO_4 \cdot PbO$ or $(PbSO_4)_2 \cdot Pb(OH)_2$. *Chrome yellows and oranges* consist of various proportions of lead chromate, $PbCrO_4$ (the chief constituent), lead sulfate, $PbSO_4$, and lead monoxide, PbO. *Chrome greens* are mixtures of Prussian blue, $Fe_4[Fe(CN)_6]_3$, and lead chromate, $PbCrO_4$. *Molybdate oranges* with a red tendency are made from lead chromate precipitated together with lead molybdate, $PbMoO_4$, and lead sulfate, $PbSO_4$. *Red lead*, Pb_3O_4, is used extensively as a primer paint.

A commonly held belief is that lead-containing pigments react with hydrogen sulfide in polluted air to form the black precipitate lead sulfide. On exposure to hydrogen sulfide gas, all of these pigments will darken because of the formation of the black lead sulfide, PbS:

$$PbCO_3 \text{ (white)} + H_2S \rightarrow PbS \text{ (black)} + H_2O + CO_2.$$

However, hydrogen sulfide in industrial atmospheres is rapidly oxidized to sulfur dioxide and sulfuric acid.

Alternative explanations for the discoloration of old paintings are available. For example, exposure of lead-based pigments to sulfide-based pigments will also contribute to the darkening over time. Specifically, white lead and chrome yellow (lead-containing pigments) should not be used with the blue pigment *ultramarine*, a complex sodium aluminum silicate and sulfide found in the mineral *lapis lazuli*. The key ingredients for the formation of black lead sulfides are present when combinations of these dyes are used.

Other chemical factors can cause pigments to darken. Cuprous oxide, Cu_2O, starts out as bright red pigment, but it gradually oxidizes to the cupric form, CuO, which is characteristically black. Prolonged exposure to light can cause copper resinate, a transparent green pigment used in the fifteenth–eighteenth centuries, to become a deep chocolate brown. Prussian blue, $Fe_4[Fe(CN)_6]_3$ or ferric ferrocyanide, should not be used with basic carbonate white lead or other basic pigments because of the reaction to precipitate ferric hydroxide, imparting a reddish tinge to the paint:[1]

$$Fe^{3+} \text{ (aq)} + 3 \, OH^- \text{ (aq)} \rightarrow Fe(OH)_3 \text{ (s)}.$$

KEY TERMS: **oxidation** **precipitation**

References

[1] "Dosimetry of the Museum Environment: Environmental Effects on the Chemistry of Paintings." Oscar F. van den Brink, Molecular Aspects of Ageing in Painted Works of

Art, Progress Report 1995–1997, FOM Institute for Atomic and Molecular Physics, Amsterdam, NL, http://www.amolf.nl/research/biomacromolecular_mass_spectrometry/molart/progress_report98.pdf

[2] "Molecular Aspects of Ageing in Painted Works of Art." Progress Report 1995–1997, FOM Institute for Atomic and Molecular Physics, Amsterdam, NL, http://www.amolf.nl/research/biomacromolecular_mass_spectrometry/molart/progress_report98.pdf

[3] K. K. Stevens, and J. C. Warner, "Organic Protective Coatings: Paints, Varnishes, Enamels, Lacquers," in *Chemistry of Engineering Materials*, fourth ed., ed. J. C. Warner (New York: McGraw-Hill, 1953), 544–590.

Related Web Sites

▶ "An Investigation of Paul Cezanne's Watercolors with Emphasis on Emerald Green." Faith Zieske, Philadelphia Museum of Art, The Book and Paper Group Annual, Volume 14, 1995, http://aic.stanford.edu/conspec/bpg/annual/v14/bp14-09.html

▶ "A Team from the Universitat Politecnica de Catalunya Verify the Authenticity of a Sketch by Francisco de Goya Recently Found in Palafrugell (Girona)." Universitat Politecnica de Catalunya, Archive, Past Research and Institutional News: Materials Technology News, http://www.upc.es/op/english/noticies/acresearch/1996/Goyaauthentic.html

▶ "Yellow Lake Pigments, Identification and Study of Deterioration of Flavonoid Colorants." Arie Wallert and N. Wyplosz, Molecular Aspects of Ageing in Painted Works of Art, Progress Report 1995–1997, FOM Institute for Atomic and Molecular Physics, Amsterdam, NL, http://www.amolf.nl/research/biomacromolecular_mass_spectrometry/molart/progress_report98.pdf

▶ "Orpiment, Deterioration of Arsenic Sulfide Pigments." Arie Wallert, Molecular Aspects of Ageing in Painted Works of Art, Progress Report 1995–1997, FOM Institute for Atomic and Molecular Physics, Amsterdam, NL, http://www.amolf.nl/research/biomacromolecular_mass_spectrometry/molart/progress_report98.pdf

▶ "Cadmium Pigments." J. R. J. van Asperen de Boer, Molecular Aspects of Ageing in Painted Works of Art, Progress Report 1995–1997, FOM Institute for Atomic and Molecular Physics, Amsterdam, NL, http://www.amolf.nl/research/biomacromolecular_mass_spectrometry/molart/progress_report98.pdf

5.14 How Do Hair Coloring Products "Remove" Gray in Hair?

Grecian Formula works with the hair's natural chemistry to produce a similar acting pigment where your hairs' melanin used to be. Grecian Formula combines with the hair's protein inside the hair to gradually "reverse" the graying process and bring back natural looking color.

—Grecian Formula 16, packaging copy

The Chemical Basics

Many hair-coloring products react with the proteins in hair to produce a black-colored compound that "covers" the gray.

The Chemical Details

Some hair-coloring products contain the water-soluble compound lead (II) acetate, $Pb(CH_3COO)_2$. When the coloring product is applied to the hair, a chemical reaction occurs between the Pb^{2+} ion and the sulfur atoms in cysteine and methionine incorporated in amino acids in hair proteins. The insoluble black product lead(II) sulfide forms.

> **KEY TERMS:** solubility   precipitation

Related Web Sites

▶ "What's That Stuff?" Linda Raber, *Chemical and Engineering News OnLine* **78** March 13, 2000, Volume 78, Number 11, p. 52, http://pubs.acs.org/cen/whatstuff/stuff/7811scit4.html

▶ "How Hair Coloring Works." Debbie Selinsky
   http://www.howstuffworks.com/hair-coloring.htm

5.15 What Is the Source of Dr. Seuss' Green Eggs (and Ham)?

Say! I like green eggs and ham! I do! I like them, Sam-I-am!

—Dr. Seuss, *Green Eggs and Ham*

The Chemical Basics

Eggs that are cooked too long or cooled too quickly can develop green yolks or concentric rings of a greenish-tinge in the yolk. While this greenish color is harmless, the green appearance is a consequence of a reaction in the egg to produce a compound known as ferrous sulfide or iron(II) sulfide. An egg is a good source of both iron and protein. A typical egg contains 0.590 mg iron in the yolk.[1] Heating an egg to temperatures above $70°C$[2,3] can cause the sulfur-containing amino acids in the proteins to decompose and generate hydrogen sulfide. Combination of the yolk iron and the sulfide decomposition product will produce the green iron(II) sulfide. The concentric rings reflect the development of a hen's yolk in spherical layers and variations in the iron content of the hen's feed or water as the egg yolk grows.[2] The water used to hard boil an egg can also be an additional source of iron.

The Chemical Details

Methionine (Fig. 5.15.1) and cysteine (Fig. 5.15.2) are the sulfur-containing amino

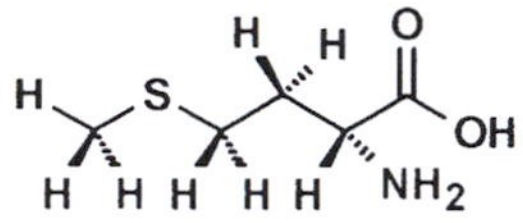

Figure 5.15.1 ▶ The molecular structure of methionine.

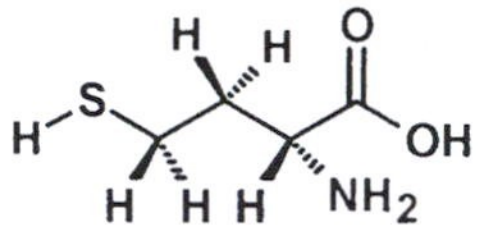

Figure 5.15.2 ▶ The molecular structure of cysteine.

acids.

    Thermal degradation of these amino acids yield such end products as hydrogen sulfide (H_2S), ammonia (NH_3), and methane (CH_4), and organic acids such as acetic acid (CH_3COOH) and formic acid ($HCOOH$).

KEY TERMS: precipitation   solubility   amino acids

References

[1] "Free Range Egg Facts." British Free Range Egg Producers Association, http://www.bfrepa.co.uk/eggfacts.htm

[2] "The Science of Boiling an Egg." Charles D. H. Williams, University of Exeter, School of Physics, http://newton.ex.ac.uk/teaching/CDHW/egg/#hard

[3] "Nutrition and Food Management 236: Egg." Oregon State University http://osu.orst.edu/instruct/nfm236/egg/index.cfm#e

Related Web Sites

▶ "Food Product Design: Protein Possibilities." Lynn A. Kuntz, October 1997, http://www.foodproductdesign.com/archive/1997/1097AP.html

5.16 How Does Washing Soda Soften Hard Water?

The addition of washing soda to water has been recommended to accomplish a number of tasks, including softening hard water as well as improving the cleaning ability of a detergent. A simple chemical reaction contributes to the effectiveness of this substance.

The Chemical Basics

Water containing high concentrations of dissolved solids is often described as hard water. Calcium and magnesium salts are the common salts dissolved in ground water as a result of a partial dissolution of certain minerals. Limestone, dolomite, and gypsum are the main sources of these ions. One way to remove the dissolved calcium and magnesium ions is to add a substance that will combine with these ions to form a solid precipitate that can be filtered from the water. Washing soda is a solid substance that dissolves in water to release carbonate ions that subsequently react with calcium and magnesium ions to form an insoluble substance.

The Chemical Details

Limestone is sedimentary rock composed mainly of calcium carbonate, usually in the form of calcite (the rhombohedral crystal structure) or aragonite. The mineral dolomite is also sedimentary rock that is often described as calcium magnesium carbonate, $CaMg(CO_3)_2$, where magnesium ions substitute for calcium ions within the crystal structure. Gypsum is a mineral composed of hydrated calcium sulfate ($CaSO_4 \cdot 2\,H_2O$). The action of rainwater on limestone is a common mechanism for the generation of calcium ions in groundwater. While calcium carbonate is fairly insoluble in water, the dissolution of limestone is aided by the added acidity of rainwater. Rain water absorbs carbon dioxide from the atmosphere, soil, and decaying vegetation to form a weak carbonic acid solution:

$$H_2O\ (l) + CO_2\ (g) \rightleftarrows H_2CO_3\ (aq)$$
$$H_2CO_3\ (aq) + CaCO_3\ (s) \rightarrow Ca(HCO_3)_2\ (aq).$$

The action of carbonic acid on limestone produces a calcium bicarbonate solution that is exceedingly soluble in water. (For comparison, at $20°C$ the solubility of calcium carbonate in water is only 0.0145 g per liter while the solubility of calcium bicarbonate is 166 g per liter.[1]) Magnesium ions from dolomite are also released into aqueous solution according to the same mechanism. The weathering of gypsum, calcium sulfate, also releases calcium ions into natural water supplies.

One way that the cations calcium and magnesium may be removed is by adding "washing soda," sodium carbonate, $Na_2CO_3 \cdot 10\,H_2O$. This sodium salt is highly soluble in aqueous solution. The process of adding washing soda to hard water is aimed at precipitating the calcium and magnesium cations with a single anion via the salts with the lowest combination of solubility products. With the solubility product values indicated below, the solubility of each carbonate can be easily calculated:

$$CaCO_3 \quad K_{sp} = [Ca^{2+}][CO_3^{2-}] = 8.7 \times 10^{-9}$$
$$S \cdot S = 8.7 \times 10^{-9}$$
$$S = 9.3 \times 10^{-5} \text{ M}$$
$$MgCO_3 \quad K_{sp} = [Mg^{2+}][CO_3^{2-}] = 4.0 \times 10^{-5}$$
$$S \cdot S = 4.0 \times 10^{-5}$$
$$S = 6.3 \times 10^{-3} \text{ M}$$

If S moles of $CaCO_3$ dissolve in a liter of water, then S moles each of calcium ion and carbonate ion form. With these ion concentrations equal to S, the solubility of $CaCO_3$ is calculated as 9.3×10^{-5} M. The higher solubility of magnesium carbonate in water, 6.3×10^{-3} M, results from the larger solubility product constant. Nevertheless, both of these carbonate salts are rather insoluble, and the excess carbonate anions provided by the sodium carbonate effectively precipitate the calcium and magnesium ions from solution.

KEY TERMS: precipitation   solubility

References

[1] "The Soil as an Environment for Microorganisms." N. A. Krasil'nikov, Academy of Sciences of the USSR, Institute of Microbiology, http://www.soilandhealth.org/01aglibrary/010112Krasil/010112krasil.ptII.html

Related Web Sites

- "Hardness of Water." http://www.mp-docker.demon.co.uk/environmental_ chemistry/topic_3b/index.html

- "Water Quality Constituents and Their Significance." Montana Bureau of Mines and Geology: Ground-Water Information Center, http://mbmggwic.mtech.edu/tools/dict/qw_table1.html

- "The Chemistry of Natural Waters." University of Wisconsin–Superior, http://acad.uwsuper.edu/uwsscied/Life/LABS/H2Ochem.htm

- "The Calcium Cycle." Bill Duesing, Yale-New Haven Teachers Institute, http://www.cis.yale.edu/ynhti/curriculum/units/1985/7/85.07.08.x.html

THERMAL DECOMPOSITION

5.17   Why Does Baking Soda Extinguish a Fire?

A box of baking soda is often recommended as a handy fire extinguisher in the kitchen. This household hint takes advantage of a chemical reaction involving baking soda as a reactant to douse the fire.

The Chemical Basics

Three key components are required to sustain a fire: a fuel (e.g., a hydrocarbon compound such as methane or octane or a piece of wood containing cellulose), an oxidizing agent (generally supplied by oxygen in air), and heat. Combustion of the fuel is terminated when the supply of either the fuel, the oxidizing agent, or the heat is eliminated. Many firefighting techniques focus on removing the last two ingredients. For example, fires are often smothered by restricting the flow of oxygen to the fuel, as in the action of extinguishing a flame by throwing a fire blanket over the burning object. Similarly, the oxygen atmosphere can be replaced by a gas that will not support combustion, such as carbon dioxide (CO_2). The action of CO_2 fire extinguishers relies on the fact that CO_2 is heavier than air (44.0 vs 28.9 g mol^{-1}),[1] enabling a CO_2-enriched environment to settle over and envelop the burning fuel. The displacement of air by the denser CO_2 gas terminates the combustion process. Finally, water battles fires by cooling the system and lowering the temperature to a point where the reaction known as burning ("combustion") can no longer be sustained.

Solid sodium bicarbonate, $NaHCO_3$, otherwise known as baking soda, is also an excellent fire extinguisher because the chemistry of this salt at high temperatures diminishes the oxygen environment. How does this occur? While $NaHCO_3$ is a stable ionic solid at room temperature, at the high temperatures typical of fires $NaHCO_3$ (s) undergoes *thermal decomposition* (i.e., a heat-activated reaction) to produce carbon dioxide gas as one of the by-products. Just as in the case of CO_2 fire extinguishers, the CO_2 gas produced by the decomposition of baking soda creates an atmosphere of reduced oxygen content that aids in extinguishing a fire.

The Chemical Details

A thermal decomposition reaction is a reaction that is activated by heat or high temperatures and that generates simpler (i.e., containing fewer atoms and thus characterized by lower molecular weights) substances from a single complex substance. The overall balanced equation for the thermal decomposition of sodium bicarbonate reveals the simpler substances produced:

$$2 \, NaHCO_3 \, (s) \rightarrow Na_2CO_3 \, (s) + H_2O \, (g) + CO_2 \, (g).$$

In addition to carbon dioxide gas and water vapor, a white crystalline solid, sodium carbonate, is formed. How does heat induce this reaction? The requirement of an input of heat to decompose the ionic lattice of sodium bicarbonate is supported by a calculation[2] of the standard enthalpy change for this reaction at 298 K, an endothermic quantity of $+135.6$ kJ mol^{-1}, meaning heat is required for the reaction to occur. The overall balanced equation also clearly depicts the heterogeneous nature of the decomposition process. A *heterogeneous reaction* is characterized by the presence of multiple phases in the reaction mixture; here both solid and gas phases exist at the high temperatures at which this reaction occurs. One of the driving forces for this reaction to go to completion is the loss of the gas products to the atmosphere. Thus, the chemistry of sodium bicarbonate at elevated temperatures is a convenient source of carbon dioxide to eliminate the flow of oxygen to a burning fuel.

KEY TERMS:	heterogeneous reaction	combustion
	thermal decomposition reaction	fuel
	endothermic reaction	oxidizing agent

References

[1] Using a composition for dry air of approximately 78.1% N_2 (g), 21.0% O_2 (g), and 0.9% Ar (g), an average molecular weight for air is estimated at 28.9 g mol^{-1}.

[2] Using standard molar enthalpies of formation at 298 K for NaHCO$_3$ (s), Na$_2$CO$_3$ (s), H$_2$O (g), and CO$_2$ (g) of -950.8, -1130.7, -241.8, and -393.5 kJ mol^{-1}, respectively. (*CRC Handbook of Chemistry and Physics*, 74th ed., **5**-4–**5**-47.)

Related Web Sites

▶ "Arm & Hammer Baking Soda: The Everyday Miracle." http://www.armandhammer.com/FrontPorch/fs.htm

---

5.18 What Is the Origin of the Expression "in the Limelight"?

We often use the expression "in the limelight" to characterize someone at the center of attention. Theatrical production in the mid-nineteenth century created "limelight" through a chemical reaction to illuminate the stage and assist theater-goers in viewing the star performers.

The Chemical Basics

Limelights were a lighting system invented by the British engineer Captain Thomas Drummond in 1816 to use for surveying purposes. These novel lighting systems were essentially very bright gas lamps that used the heated element rather than the gas flame to generate light. When heated in a flame consisting of jets of oxygen and hydrogen gas (an oxyhydrogen flame), a block of lime (calcium oxide) becomes incandescent and emits a soft, brilliant white light. (*Incandescence* is a general term to describe light produced by heating a solid. Incandescent light bulbs heat solid tungsten filaments to produce light.) While the sharp point of an oxyhydrogen flame generates a small area of incandescence, mirrored reflectors can be used to direct the intense light and expand the area of illumination. Limelights were first used in the theatre in 1837 and were widely employed by the 1860's.[1] The visibility of the light over long distances (over 66 miles according to Drummond)[2] led to the use of limelight in lighthouses. For example, in 1861 the illuminating properties of the limelight were tested in the lighthouse on the white chalk cliffs of the South Foreland, the closest point of approach of mainland England to France.[3] The ability to achieve remote lighting of the stage with a powerful source that could be focused and varied in brightness for special effects largely contributed to the popularity of this lighting system. Illumination of front and center stage by limelights in the form of "spotlightling" led to the reference of the most desirable acting area on stage as "in the limelight." As the lime is consumed by burning, continual illumination requires an operator to constantly supply the flame with a fresh surface of lime, a drawback of this lighting method.

The Chemical Details

Calcium oxide can be produced from extensive heating of limestone. Primarily composed of calcium carbonate, limestone is extracted from both underground and surface mines and heated to temperatures exceeding 180°F to convert the calcium carbonate into calcium oxide. This thermal decomposition reaction also generates carbon dioxide gas.

$$CaCO_3 \text{ (s)} \rightarrow CaO \text{ (s)} + CO_2 \text{ (g)}.$$

Calcium oxide crystallizes in the sodium chloride structure, a structure with two interpenetrating face-centered cubic lattices. In the NaCl lattice depicted in Fig. 5.18.1, the face-centered cubic arrangement of sodium cations (the smaller spheres) is readily apparent with the larger spheres (representing chloride anions) filling what are known as the *octahedral holes* of the lattice. Octahedral holes are defined as cavities in a crystal lattice that have six identical and equidistant atoms or ions as the nearest-neighboring species. The NaCl structure is also characterized by an arrangement that allows each cation to have six equidistant anions as nearest-neighboring ions and each anion to have six equidistant neighboring cations. Thus, both the sodium and chloride ions are said to have a *coordination number* of six.

The chemistry of calcium oxide limits the lifetime of the limelight. Exposure at ordinary temperatures to water moisture and carbon dioxide in the atmosphere

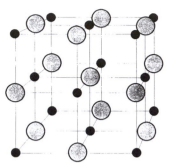

Figure 5.18.1 ▶ The NaCl crystal structure consisting of two interpenetrating face-centered cubic lattices. The face-centered cubic arrangement of sodium cations (the smaller spheres) is readily apparent with the larger spheres (representing chloride anions) filling what are known as the *octahedral holes* of the lattice. Calcium oxide also crystallizes in the sodium chloride structure.

ultimately yields calcium carbonate, which fractures upon strong heating.

$$CaO\ (s) + H_2O \rightarrow Ca(OH)_2\ (s)$$
$$Ca(OH)_2\ (s) + CO_2\ (g) \rightarrow CaCO_3\ (s) + H_2O\ (l)$$
$$CaO\ (s) + CO_2\ (g) \rightarrow CaCO_3\ (s).$$

It was common practice in the nineteenth century to wrap the lime block in dense paper or with a coating of wax to prolong the use of the lighting element.[3] Storage of the lime in a sealed can to limit contact with air also extended its utility.

KEY TERMS:	face-centered cubic crystal lattice	octahedral hole
	thermal decomposition	coordination number
	lime	

References

[1] "The Limelight." Tulsa Stage Employees, IATSE Local 354, http://www.iatse354.com/354/354html/limelight.htm

[2] M. B. Hocking, and M. L. Lambert, "A Reacquaintance with the Limelight." *Journal of Chemical Education* **64** (1987), 306–310.

[3] "World Lighthouse Information: The Lighthouses of England: South Foreland Contents: Experiments in Illumination." http://www.btinternet.com/~k.trethewey/

Related Web Sites

▶ "Lime: The Essential Chemical." The National Lime Association, http://www.lime.org/aboutlime.html

▶ "Light Producing Technologies." Division of Informatics, University of Edinburgh
http://www.dai.ed.ac.uk/CVonline/LOCAL_COPIES/RUSK/Light
Technologies.html

5.19 What Causes Puff Pastry to Expand?

Who hasn't been tantalized and impressed by the delectable, eye-catching specialities prepared with puff pastry: Chocolate napoleans, Beef Wellington, baked brie, pate en croute, etc. While the secret of these culinary delights may seem more of an art than a science, professional pastry chefs know the chemistry necessary to create the flaky, tender pastry for these delicacies.

The Chemical Basics

The baking process creates the flaky layers in puff pastries (*paté feuilletee*) used for rich napoleon desserts and chicken *vol-au-vents* and generates the air pockets in chou pastes that are filled to create cream puffs and eclairs. Heat alone will not achieve the desired effect. Substances known as *leavening agents* cause the expansion of doughs and batters through the release of gases in these baking mixtures. Common leavening agents include substances such as yeast, baking soda, and baking powder. However, a leavening effect can also be achieved by entrapping air in the batter through vigorous beating ("air leavening," as in angel food cakes and sponge cakes) and by the vaporization of volatile liquids due to heat from an oven. When the volatile liquid is water, the process is referred to as steam leavening. Water vapor pressure is fairly insignificant at room temperature but rises substantially as the boiling point of water is approached. The volume expansion of steam creates the puffed nature of the pastry or the interior cavities for cream puffs and eclairs.

The Chemical Details

What are the gases produced from the heating of leavening agents? When the leavening agent is baking soda or sodium bicarbonate ($NaHCO_3$), the gas carbon dioxide (CO_2) is released when the baking soda combines with an acidic ingredient in the recipe:

$$NaHCO_3 \text{ (s)} + H^+ \text{ (aq)} \rightarrow Na^+ \text{ (aq)} + H_2O \text{ (l)} + CO_2 \text{ (g)}.$$

Common acidic ingredients include vinegar, lemon juice, sour milk, buttermilk, yogurt, tart fruits, and cream of tartar. Commercial bakeries often use ammonium bicarbonate or ammonium carbonate as a leavening agent. The gas-producing reaction with ammonium bicarbonate actually generates both carbon dioxide gas and ammonia gas:

$$NH_4HCO_3 \text{ (s)} \rightarrow NH_3 \text{ (g)} + H_2O \text{ (l)} + CO_2 \text{ (g)}.$$

The distinctive aroma of ammonia is often apparent in bakeries but not in the final product. Bakers' yeast performs its leavening function by fermenting such sugars as glucose, fructose, maltose, and sucrose. The principal products of the fermentation process are carbon dioxide gas and ethanol, an important component of the aroma of freshly baked bread. The fermentation of the sugar, glucose — an example of a decomposition reaction — is given by the equation in Fig. 5.19.1.

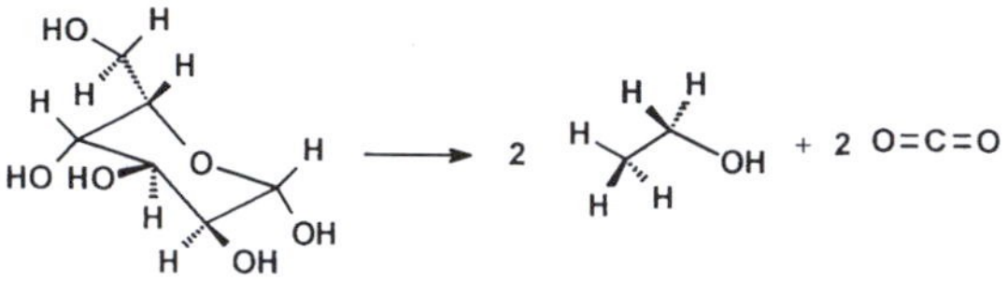

Figure 5.19.1 ▶ The fermentation of glucose to yield ethanol and carbon dioxide (a decomposition reaction).

The key to the action of steam leavening is the temperature-dependent vapor pressure of water. The vapor pressure of a liquid at a given temperature is the pressure of the gas in equilibrium with the liquid. The liquid–gas equilibrium is characterized by equal rates of vaporization and condensation at the molecular level (a "dynamic" equilibrium) and a constant or "equilibrium" vapor pressure at the macroscopic level. Two key factors affect a liquid's vapor pressure — the intermolecular forces exerted between molecules and the liquid's temperature. The stronger the intermolecular forces, the less likely the liquid is to escape to the gas phase, thus keeping the vapor pressure low. The higher the temperature, the greater the motion of the molecules and the more likely the molecules are to overcome the existing intermolecular forces, increasing the vapor pressure. The dependence of the vapor pressure of water on temperature is illustrated in Fig. 5.19.2. The curve on the left is generated by plotting vapor pressure directly as a function of temperature. On the right, a linear relationship is obtained by graphing ln P as a function of the reciprocal of temperature. This relationship is expressed mathematically in the form of the *Clausius–Clapeyron equation*

$$\ln P = \frac{-\Delta H_{vap}}{R}\left(\frac{1}{T}\right) + \frac{\Delta S_{vap}}{R}.$$

The slope of the line allows for the determination of the enthalpy of vaporization of water, ΔH_{vap}, and the y intercept yields the entropy of vaporization, ΔS_{vap}. As both the enthalpy and the entropy of water increase as the phase change *liquid* → *vapor* occurs, the slope and y intercept of the Clausius–Clapeyron equation are negative and positive, respectively. At 373 K these thermodynamic quantities have values of ΔH_{vap} = 40.657 kJ mol^{-1} [1] and ΔS_{vap} = 109.0 J K^{-1} mol^{-1}.[2]

The leavening action due to water vapor or steam arises from the increased amount of water vapor that forms as pastry temperatures initially rise in the oven and then from the increased volume of the water vapor as temperatures continue

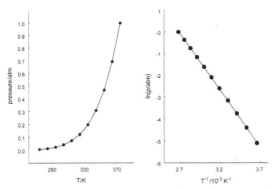

Figure 5.19.2 ► A graph of the vapor pressure of liquid water as a function of temperature (on the left), and a graph of the natural logarithm of the vapor pressure of water as a function of the reciprocal of temperature.

to rise to the desired baking temperature. Let's look at this volume expansion in quantitative detail. First of all, how do the volumes of one mole of liquid water and one mole of liquid vapor at 373 K and 1 atm compare? The volume occupied by one mole or 18.02 g of liquid water under these conditions is equal to the quantity *weight/volume* or $(18.02 \text{ g mol}^{-1})/(0.95840 \text{ g cm}^3) = 18.80 \text{ cm}^3 = 18.80 \text{ mL}$.[1] Using the ideal gas law, $PV = nRT$, the corresponding amount of water vapor occupies $(1 \text{ mol})(0.08206 \text{ L atm mol}^{-1} \text{ K}^{-1})(373 \text{ K})/(1 \text{ atm}) = 30.6$ L, a factor of 1630 times larger!

The ideal gas law further reveals that the greater the number of vapor molecules at a given temperature, the greater the volume exhibited by the gaseous phase. So as batter in an oven increases in temperature to 373 K, liquid water molecules undergo a phase transition to water gas or vapor, increasing the number of gas molecules and hence the volume of space occupied by the now gaseous water. At most baking temperatures, all water present will be in the vapor phase. As the temperature of dough rises, a greater volume is exhibited by the water vapor, again in accord with the ideal gas law. A baking temperature of 400°F (204°C or 477 K) induces a 1.3-fold increase in water vapor volume (477 K/373 K) compared with 100°C (212°F or 373 K) conditions. While water vapor can escape from the pastry, the volume expansion that occurs prior to "escape" produces the leavening action that leads to flaky pastry layers or even large void cavities.

KEY TERMS:	vapor pressure	enthalpy of vaporization
	leavening agent	entropy of vaporization
	ideal gas law	Clausius–Clapeyron equation

References

[1] D. R. Lide, ed., *Handbook of Chemistry and Physics*, 74th ed. (Boca Raton, FL: CRC Press, 1993), **6**–10.

[2] P. Atkins, *Physical Chemistry*, fifth ed. (New York: W. H. Freeman, 1994), C17.

Related Web Sites

▶ "Leavening." http://users.rcn.com/sue.interport/food/leavning.html

▶ "Baking Soda vs. Baking Powder."
http://www.users.interport.net/~sue/food/bakgsoda.html

PHOTODECOMPOSITION

5.20 Why Does Chlorine in Swimming Pools Work Best at Night?

One swimming pool dealer[1] makes the following claim: "Timing is everything — if you want an easy pool! Adding chlorine to your swimming pool in the evening, instead of the morning can cut your chemical costs in half." What valuable swimming pool chemistry is the basis for this assertion?

The Chemical Basics

Chlorine, or more commonly a substance containing hypochlorite ion, is added to pools as a disinfectant. However, sunlight rapidly destroys hypochlorite, drastically reducing the effectiveness of the sanitizer. Hence, the effectiveness of the disinfectant is maximized when added in the evening hours.

The Chemical Details

Pool chlorine, generally added in the form of calcium hypochlorite ($Ca(OCl)_2$) or sodium hypochlorite (NaOCl), readily ionizes in water to yield the hypochlorite ion, OCl^-:

$$Ca(OCl)_2 \text{ (aq)} \rightarrow Ca^{2+} \text{ (aq)} + 2\,OCl^- \text{ (aq)}$$
$$NaOCl \text{ (aq)} \rightarrow Na^+ \text{ (aq)} + OCl^- \text{ (aq)}.$$

The hypochlorite ion, a weak base, can react further with water to generate hypochlorous acid (HOCl) and hydroxide ions (OH^-) that raise the pH of the pool water:

$$OCl^- \text{ (aq)} + H_2O \rightarrow HOCl \text{ (aq)} + OH^- \text{ (aq)}.$$

Alternatively, upon absorption of ultraviolet light, a photochemical reaction can occur — that is, a reaction triggered by the energy provided from light. This reaction destroys the hypochlorite ion and produces chloride ion and oxygen gas, which escapes from the pool:

$$OCl^- \rightarrow Cl^- + \frac{1}{2}O_2 \text{ (g)}.$$

This reaction is also an oxidation–reduction process whereby the oxygen atom is oxidized from the -2 oxidation state to the zero oxidation state as the chlorine atom is reduced from the $+1$ to -1 oxidation state. As diatomic oxygen is an effective disinfectant, pool owners should avoid the loss of O_2 via the decomposition of the hypochlorite ion. Adding hypochlorite-containing disinfectant in the evening hours reduces the loss of the ion from photochemical decomposition.

KEY TERMS: oxidation–reduction photochemical reaction

References

[1] "Swimming Pool Secret #6." Pool Solutions, WaterCare, Inc.,
http://www.poolsolutions.com/tip06.html

Related Web Sites

▶ "Pool Water Chemistry: Technical Details." Virtual Pool and Spa Store,
http://www.poolandspa.com/page371.htm

Other Questions to Consider

5.4   Why is hydrogen peroxide kept in dark plastic bottles? *See* p. 40.

Chapter 6 ▶

Connections to Acids and Bases

6.1 ## What Does pH Stand For?

There are countless situations where the relative acidity or alkalinity of a substance or system is of critical importance. Agricultural conditions, water quality, food preservation, lung and kidney function—all are circumstances in which quantitative measures of acidity or basicity enable proper maintenance and regulation of vital processes. The term pH is widely used as an expression of acid/base content, but what is the origin of such nomenclature?

The Chemical Basics

The pH of a substance, usually a solution, is a quantitative measure of the acidity of the substance. The Swedish chemist Svante Arrhenius proposed in the late 1880s that acids were substances that delivered hydrogen ions to a solution. In 1904 H. Friedenthal recommended that acidity be expressed in terms of the concentration of the hydrogen ion present. In 1909 the Danish biochemist Soren P. L. Sorenson published his classic paper, "Enzyme Studies II. The Measurement and Meaning of Hydrogen Ion Concentration in Enzymatic Processes," in the journal *Biochemische Zeitschrift*.[1] In this work he introduced a quantitative expression for the acidity of a solution. Interestingly, his designation for the parameter was written PH, not pH. Mathematically speaking, Sorenson defined pH as the negative of the logarithm (base 10) of the hydrogen ion concentration expressed as a power of 10. Operationally, Sorenson showed that the pH of a solution can be easily expressed in a manner that reveals the relative acidity of the system through *the power of the hydrogen* ion concentration (or pH):

> The value of the hydrogen ion concentration will accordingly be expressed by the hydrogen ion..., and the factor will have the form of

a *negative power of* 10... I will explain here that *I use the name "hy-drogen ion exponent" and the designation P_H for the numerical value of the exponents of this power.*[1]

The Chemical Details

Sorenson chose to write the concentration of hydrogen ions as a power of 10. For example, an aqueous solution that contained 5.00×10^{-2} M hydrogen ions (H^+) would be denoted as containing the mathematically equivalent value of $10^{-1.30}$ M H^+ ions. Sorenson designated the "P_H" or pH of the solution as the numerical value of the negative exponent of 10. Thus, a pH of 1.30 would be ascribed to the solution. In other words,

$$[H^+] = 10^{-pH}.$$

The concept of using the base 10 logarithm to express the magnitude is a widespread practice today. Equilibrium constants of chemical reactions are often noted or compared as pK values where p$K = -\log 10$ (magnitude of equilibrium constant). For example, the extent of dissociation of acetic acid, the acid in vinegar, is quantified by an equilibrium constant of 1.8×10^{-5}. Here, then, p$K = -\log_{10}(1.8 \times 10^{-5}) = 4.74$.

KEY TERMS: pH   logarithm   acidity   concentration

References

[1] "Classic Papers in Chemistry": S. P. L. Sorenson, "Enzyme Studies II. The Measurement and Meaning of Hydrogen Ion Concentration in Enzymatic Processes." *Biochemische Zeitschrift* **21,** (1909) 131–200,
http://dbhs.wvusd.k12.ca.us/Chem-History/Sorenson-article.html

Related Web Sites

► "Soren Sorenson and the pH Scale." John L. Park,
http://dbhs.wvusd.k12.ca.us/AcidBase/pH.html

► "pH Values of Various Foods." U.S. Food & Drug Administration, Center for Food Safety & Applied Nutrition, http://vm.cfsan.fda.gov/~mow/app3a.html

6.2   Why Does Disappearing Ink Disappear?

Disappearing ink or invisible ink — chances are, you've been fascinated by the gradual fading of color when a message written with seemingly standard ink dissipates right before your eyes. Why should a chemist not be fooled by the disappearing ink in a magician's bag of tricks?

The Chemical Basics

Whether for a class demonstration, a practical joke, or perhaps a clandestine activity, disappearing ink is a fascinating substance. What is the secret to its action? One formulation of disappearing ink contains a common acid–base indicator, that is, a substance that by its color shows the acid or basic nature of a solution. One acid–base indicator that shifts from a colorless hue under acidic conditions to a deep blue color in alkaline solutions is thymolphthalein. If the indicator starts off in a basic solution, perhaps containing sodium hydroxide, the typical blue color of an ink is perceived. How does the ink color disappear? This behavior is dependent upon the contact of the ink with air. Over time, carbon dioxide in the air combines with the sodium hydroxide in the ink solution to form a less basic substance, sodium carbonate. The carbon dioxide also combines with water in the ink to form carbonic acid. The indicator solution responds to the production of acid and returns to its colorless acid form. A white residue (sodium carbonate) remains as the ink dries.

The Chemical Details

Thymolphthalein (Fig. 6.2.1) (also known as 2′,2″-dimethyl-5,5-di-iso-propyl-

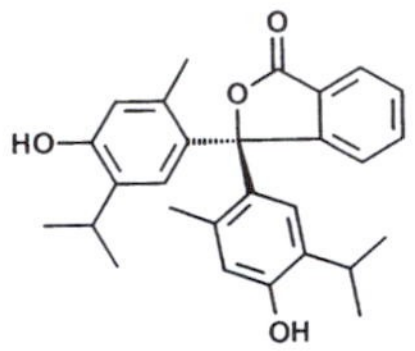

Figure 6.2.1 ▶ The molecular structure of thymolphthalein (2′,2″-dimethyl-5,5-di-iso-propylphenolphthalein) is an acid–base indicator that is colorless in its acidic form and deep blue in its basic form.

phenolphthalein) is an acid–base indicator that is colorless in its acidic form and deep blue in its basic form. The equilibrium between the acidic and basic forms of the indicator may be represented as

$$\text{HIn (colorless)} \rightleftarrows \text{H}^+ + \text{In}^- \text{ (deep blue).}$$

The acidic form of the indicator (HIn) retains the hydrogen on each hydroxyl group; the conjugate base form of the indicator (In⁻) contains one ionized hydroxyl group (−O⁻). The pK_a value for the acid ionization is 9.9; thus, $K_a = 10^{-9.9} = 1.3 \times 10^{-10}$.[1] A discernible color change is noted when the pH of an aqueous solution of the indicator is in the range of 9.4 to 10.6.

Disappearing ink can be prepared by first dissolving solid thymolphthalein in ethanol, adding water, and then adjusting the pH with sodium hydroxide solution.[2] The deep blue color of the basic form of the indicator is readily apparent. Applying the ink to paper increases its exposure to carbon dioxide in air. Two chemical reactions occur. Carbon dioxide and sodium hydroxide react to form the salt sodium carbonate:

$$2\,NaOH\,(aq) + CO_2\,(aq) \rightarrow Na_2CO_3\,(s) + H_2O\,(l).$$

Carbon dioxide and water also combine to form carbonic acid:

$$CO_2\,(aq) + H_2O\,(l) \rightarrow H_2CO_3\,(aq).$$

The partial ionization of carbonic acid produces hydronium ion, H^+, driving the indicator equilibrium to the weak acid form. A colorless solution results. As the water in the ink evaporates, the white residue of sodium carbonate remains.

KEY TERMS: acid–base indicator   conjugate base   ionization

References

[1] "Properties of Aqueous Acid-Base Indicators at 25°C." Chemical Sciences Data Tables, James A. Plambeck, 1996, http://www.compusmart.ab.ca/plambeck/che/data/p00433.htm

[2] "Thirty Demos in Fifty Minutes." Courtney W. Willis, Physics Department, University of Northern Colorado, and Meg Chaloupka, Windsor High School, http://physics.unco.edu/sced441/demos96.html

6.3   How Are Light Bulbs Frosted?

In 1925 Kyozo Fuwa (Toshiba) developed the first frosted glass light bulb, generally regarded as one of the five greatest light bulb inventions.[1] The softer light created by a frosted lightbulb is accomplished through a chemical reaction with the inner surface of the bulb.

The Chemical Basics

The interior of a "frosted" light bulb is obtained by etching with hydrofluoric acid, HF. A chemical reaction between the glass and the acid produce a white substance that coats the bulb's interior surface. Frosted light bulbs were introduced in 1925 to provide a diffused light rather than the glaring light of an unconcealed filament.[1] The decorative design of etched glass is also accomplished through the corrosive action of hydrofluoric acid. Typically, glass is coated with layers of paraffin or beeswax through which patterns are traced with metal needles. Dipping the glass in an aqueous solution of HF etches the design in the unprotected surface.

The Chemical Details

Hydrofluoric acid reacts with glass via an overall reaction that may be summarized as

$$SiO_2 \text{ (s)} + 6 \text{ HF (aq)} \rightarrow H_2SiF_6 \text{ (aq)} + 2 \text{ H}_2O \text{ (l)}.$$

(Recall that, because of the strong H–F bond, hydrofluoric acid is a weak acid with a small acid dissociation constant K_a of 6.8×10^{-4}. In contrast, the other binary acids of the halogen family — HCl, HBr, and HI — are strong acids that completely dissociate in water.) The fluorosilicic acid produced, H_2SiF_6, is a water-soluble substance with a structure as in Fig. 6.3.1.

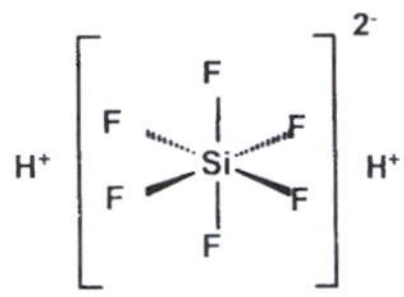

Figure 6.3.1 ▸ The molecular structure of fluorosilicic acid, produced by the action of hydrofluoric acid (HF) on glass.

As we know, glass is defined as an extended three-dimensional network of atoms that forms a solid lacking the orderly arrangement or long-range periodicity of a crystalline material. Despite the amorphous nature of glass, a definite chemical composition may be ascribed to the material. Most commercially important glasses are silica glasses with SiO_4 tetrahedra as the network building blocks. The network-forming Si^{4+} silicon ions are bonded to four oxygen atoms. These oxygen atoms fall into two classifications, depending on the oxide content of the glass. A bridging oxygen atom connects two SiO_4 tetrahedra, while a nonbridging oxygen atom is linked with a "network-modifying" cation such as a sodium ion. Silica glass (also known as vitreous silica) contains 100% bridging oxygen atoms. A variety of network-modifying cations are present in soda-lime silicate glass (with O–Na, O–Ca, O–Al, and O–Mg linkages), sodium borosilicate glass (with O–Na, O–Al, and O–B linkages), and aluminosilicate glass (with linkages of oxygen to Na, Ca, Al, Mg, and B possible). Many properties of the glass — density, viscosity, chemical durability, electrical resistivity — are related to the "connectivity" of the structure, i.e., the concentration of nonbridging oxygens, and the nature of the network-modifying ions.[2] In particular, the chemical resistance of glass is enhanced by increasing the silica content at the surface. Silica glass offers the greatest resistance. Normal weathering of glass occurs via an ion-exchange process whereby alkali ions in the glass are exchanged with hydronium ions present in water or atmospheric humidity. Further depositing of white alkali carbonates and bicarbonates on the glass surface can occur as the exchanged alkali ions react with atmospheric carbon dioxide and water. Dissolution of the entire silica network is possible when hydrofluoric acid (as well as perchloric and phosphoric acids and caustic alkalis) attack the surface of silicate glasses. The chemical durability of glass can be modified by fire-polishing (which removes alkali ions by volatilization) or by treating the glass surface with a polymeric coating.

KEY TERMS: weak acid

References

[1] "Thomas Edison 1997 Main Exhibition: Filament & Incandescent Electric Light." Tepia, http://www.tepia.or.jp/main/filainfe.htm

[2] "The Structure of Glasses." Research Group Solid State Spectroscopy and Magnetochemistry in Material Science, Technische Universitt Graz, Institut fr Physikalische und Theoretische Chemie, http://www.cis.TUGraz.at/ptc/specmag/struct/sintro.htm

Related Web Sites

▶ "The Corning Museum of Glass Museum." http://www.cmog.org/home.cfm/

▶ "Lighting Systems: Bulb Finish." http://tristate.apogee.net/lite/linbfin.htm

Other Questions to Consider

3.5　　Why is milk of magnesia an antacid? *See* p. 18

9.1　　What causes the fizz when an antacid is added to water? *See* p. 116.

9.4　　Why do hydrangeas vary in color when grown in dry versus wet regions? *See* p. 122.

13.13　　When we buy fresh fish, why does its smell indicate its freshness? *See* p. 194.

Chapter 7 ▶

Connections
to Intermolecular Forces

Why Are General Anesthetics Administered as Gases?

Many surgical procedures employ the rapid, safe, and well-controlled application of gaseous anesthetics, often mixed with oxygen. These drugs are administered to obtain various stages of consciousness, muscular relaxation, and sensory stimulation. One of the characteristics of an ideal anesthetic agent is the rapid induction of these sensations using gaseous or easily volatilized inhalants. The chemical structure of an anesthetic dictates its physical state.

The Chemical Basics

General anesthetics are drugs that are administered to depress the brain's sensory response. The effectiveness of anesthetic gases depends on their ability to directly dissolve in the bloodstream and circulate to the brain. Thus, the ability of a gas to dissolve in liquids, particularly the aqueous solution that comprises blood, is critical to the function of the anesthetic. How does the drug reach the neural tissue in the brain? An equilibrium of the inhaled gas with the lung tissue first occurs to transfer the drug through the walls of the alveoli within the lungs to the arterial blood. The drug then circulates to all tissues of the body, including the brain. A subsequent equilibrium of the anesthethic between the blood and the brain tissue determines the level of unconsciousness of the patient.

The Chemical Details

Some of the general anesthetics include ether (diethyl ether) (Fig. 7.1.1), chloroform (Fig. 7.1.2), nitrous oxide (Fig. 7.1.3), ketamine hydrochloride (Fig. 7.1.4),

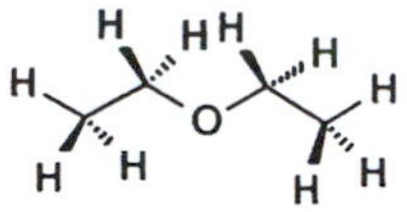

Figure 7.1.1 ▶ The molecular structure of ether (diethyl ether).

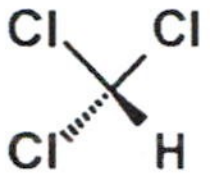

Figure 7.1.2 ▶ The molecular structure of chloroform.

and halogenated hydrocarbons such as ethyl chloride (Fig. 7.1.5), trichloroethylene (trilene) (Fig. 7.1.6), halothane (fluothane or 2-bromo-2-chloro-1,1,1-trifluoroethane) (Fig. 7.1.7), methoxyflurane (penthrane, metofane) (Fig. 7.1.8), enflurane (enthrane or 2-chloro-1,1,2-trifluoroethyl difluoromethyl ether) (Fig. 7.1.9), isoflurane (forane, aerrane, or 1-chloro-2,2,2,trifluoroethyl difluoromethyl ether) (Fig. 7.1.10), sevoflurane (ultane or 1,1,1,2,2,2-hexafluoroethyl fluoromethyl ether) (Fig. 7.1.11), and desflurane (suprane or 1,2,2,2-tetrafluoroethyl difluoromethyl ether) (Fig. 7.1.12).[1] Except for nitrous oxide, all of these substances are liquids at room temperature. Thus, the anesthetic agent must be vaporized upon administering and is usually delivered in combination with oxygen gas.

Despite the fact that the anesthetics listed above are liquids at room temperature, they all are characterized by a relative ease of volatility (i.e., tendency to vaporize or become a gas). The strength of intermolecular forces (i.e., forces between molecules) dictates the facility with which the liquid-to-gas phase transition can be accomplished. Weak dipole–dipole interactions predominate in these generally polar compounds. No hydrogen bonds can form between molecules, as no hydrogen atoms that are bonded to electronegative atoms such as oxygen, nitrogen, or fluorine are present. Thus, to change the physical state of the anesthetics from liquid to gas, only weak intermolecular forces need to be overcome. The ease of vaporization of these substances is evident in the low magnitudes of their boiling points. For example, ether has a normal boiling point of 35°C, isoflurane 48.5°C, halothane 50°C, enflurane 56.5°C, and trichloroethylene 87°C. The high volatility of these drugs is also reflected in their high vapor pressures at or near room temperature. For example, at 20°C sevoflurane has a vapor pressure of 157

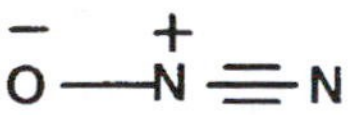

Figure 7.1.3 ▶ The molecular structure of nitrous oxide.

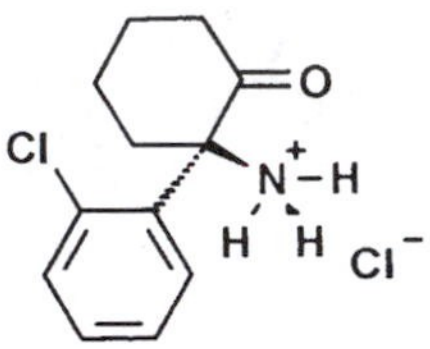

Figure 7.1.4 ►   The molecular structure of ketamine hydrochloride.

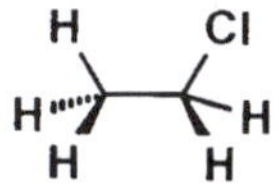

Figure 7.1.5 ►   The molecular structure of ethyl chloride.

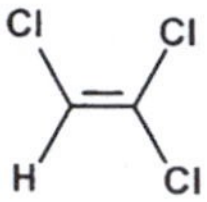

Figure 7.1.6 ►   The molecular structure of trichloroethylene (trilene).

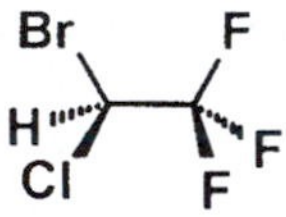

Figure 7.1.7 ►   The molecular structure of halothane (fluothane or 2-bromo-2-chloro-1,1,1- trifluoroethane).

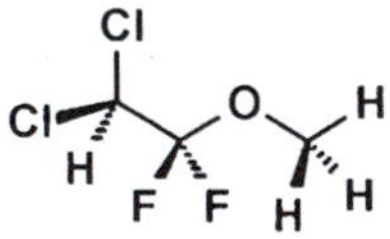

Figure 7.1.8 ►   The molecular structure of methoxyflurane (penthrane or metofane).

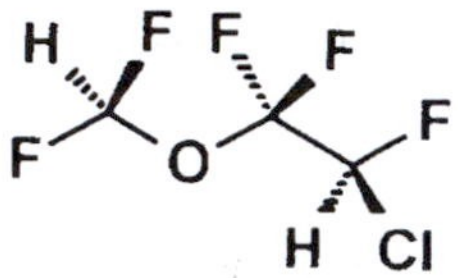

Figure   7.1.9 ►   The molecular structure of enflurane (enthrane or 2-chloro-1,1,2-trifluoroethyl difluoromethyl ether).

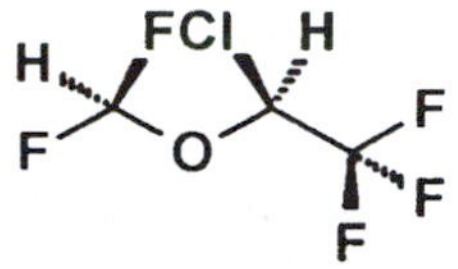

Figure 7.1.10 ▶ The molecular structure of isoflurane (forane, aerrane, or 1-chloro-2,2,2- trifluoroethyl difluoromethyl ether).

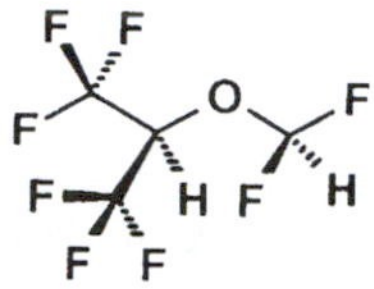

Figure 7.1.11 ▶ The molecular structure of sevoflurane (ultane or 1,1,1,2,2,2-hexafluorethyl fluoromethyl ether).

mm Hg,[2] enflurane 175 mm Hg,[3] isoflurane 238 mm Hg,[3] and desflurane 669 mm Hg.[3] For comparison, the vapor pressure of water at 25°C is 23.8 mm Hg.

The ability of the anesthetic agent to function is related to the partial pressure of the drug in the brain. Two major factors dictate the concentration of anesthetic agent in the neural tissue: (1) the pressure gradients from lung alveoli to the brain (i.e., inhaled gas → alveoli → bloodstream → brain) and (2) the lipid solubility of the drug that enables it to pass between the blood–brain barrier to the central nervous system.

The distribution of anesthetic throughout the entire body may be viewed as an equilibration process (Fig. 7.1.13), with tissues characterized by high blood flows reaching equilibration faster than muscle and fat.[4] Nevertheless, an anesthetic that is excessively soluble in blood will not partition substantially into brain and other tissues. The anesthetic properties of nitrous oxide and diethyl ether have been known since the 1840s. Zeneca Pharmaceuticals introduced the first modern inhalation anesthetic fluothane in 1957.[5] Methoxyfluorane followed in 1960, enflurane 1973, isoflurane 1981, desflurane by Anaquest (Liberty Corner, NJ) in 1992, and sevoflurane by Abbott Laboratories in 1995.[6]

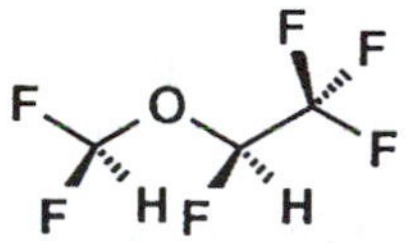

Figure 7.1.12 ▶ The molecular structure of desflurane (suprane or 1,2,2,2-tetrafluoroethyl difluoromethyl ether).

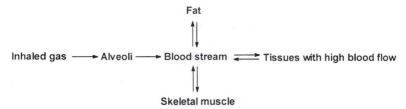

Figure 7.1.13 ▶ An overall description of the distribution of anesthetic throughout the body as an ensemble of equilibrium processes.

KEY TERMS: vapor pressure   equilibrium   vaporization
                 boiling point   volatility   intermolecular forces

References

[1] "Chemistry of General Anesthetics." PHA 422-Neurology Pharmacotherapeutics-Medicinal Chemistry Tutorial, Patrick M. Woster, Ph.D., Department of Pharmaceutical Sciences, College of Pharmacy and Allied Health Professions, Wayne State University, http://wizard.pharm.wayne.edu/medchem/ganest.html

[2] "New Anesthetic Requires New Vaporizers for Safety." William Clayton Petty, M.D., Department of Anesthesiology, Uniformed Services University of the Health Sciences, Bethesda, MD, http://gasnet.med.yale.edu/apsf/newsletter/1996/winter/apsf-new_anes.html

[3] "Anesthetic Vapor Pressure Information." Scott Medical Products, http://www.scottgas.com/medical/anestech.html

[4] "Veterinary Anesthesia: Inhalation Anesthesia." Western College of Veterinary Medicine, University of Saskatchewan, http://www.usask.ca/wcvm/anes/anesman/anes07.htm

[5] "Fluothane Anaesthetic." Zeneca Pharmaceuticals South Africa, http://home.intekom.com/pharm/zeneca/fluothan.html

[6] "Ultane (sevoflurane)." Abbott Laboratories OnLine, http://www.abbotthosp.com/PROD/anes/UltanePI.PDF

Related Web Sites

▶ "Using Volatile Anaesthetic Agents." Dr. P. Fenton, *Update in Anaesthesia, Practical Procedures*, Issue 5, 1995, http://www.nda.ox.ac.uk/wfsa/html/u05/u05_007.htm

▶ "Economic Value of Desflurane in Comparison to Isoflurane in Low Flow Anesthesia." I. Kuhn, MD, H. Wissing, MD, Department of Anesthesiology, University Hospital Frankfurt/Main, Germany, Theodor Stern Kai, D-60590 Frankfurt/Main, Germany, http://gasnet.med.yale.edu/esia/1998/may/economic.html

Why Won't Water Relieve the Burning Sensation of Chili Peppers?

Don't reach for that glass of water to cool the effect of spicy chili peppers! A few rules of chemistry will suggest a better remedy.

The Chemical Basics

Why does water "cool" the burning sensation of some "hot" (i.e., spicy) foods and not others? The cooling action arises as water acts to *dilute* the concentration of substances responsible for the burning sensation, thereby reducing their effect. To make a more dilute solution, water must be capable of dissolving the pungent ingredient. The common expression "like dissolves like" is the operational key. For any two substances, the closer their chemical structure, the more similar their intermolecular forces, that is, their attractive interactions on the molecular level. Strong intermolecular interactions lead to enhanced *solubility* (i.e., ability to dissolve). Water will cool the effect of those spicy ingredients that it can dissolve.

Substances that dissolve readily in water exhibit similar intermolecular forces to those displayed by water. In other words, compounds soluble in water are polar substances, often with the capacity to hydrogen bond to water. From your own experiences, you already recognize substances that dissolve in water and those that do not. For example, salt (ionic sodium chloride) dissolves readily in water, while oils such as butter and margarine have a limited solubility that can be enhanced by an increase in temperature. Many "hot" spices are generally nonpolar substances that dissolve only partially in water. Examples of spices with molecular structures that limit the degree of dissolution in water include zingerone, a constituent of ginger; piperine, the active component of white and black pepper; and capsaicin, the pungent ingredient in red and green chili peppers as well as in paprika.

The chemical structure of these spices renders these substances practically insoluble in water and other aqueous solutions but freely soluble in alcohols, oils, and fats (i.e., organic solvents). Hence, a drink of cold water will only temporarily relieve the burning sensation induced by the action of these ingredients on the pain-detecting nerve endings in the mouth. Alternatively, the pungency of ginger or pepper is better alleviated with a fat-based accompaniment such as sour cream or a beverage containing some alcohol.

The Chemical Details

The pungent components of chili peppers belong to a class of substances known as *capsaicinoids*. The most pungent and most common substance in this family is *capsaicin* (Fig. 7.2.1) (N-[4-hydroxy-3-methoxyphenyl)methyl]-8-methyl-6-nonenamide). Other members of this family include *dihydrocapsaicin* (Fig. 7.2.2), *nodihydrocapsaicin* (dihydrocapsaicin with a $(CH_2)_5$ linkage instead of $(CH_2)_6$),

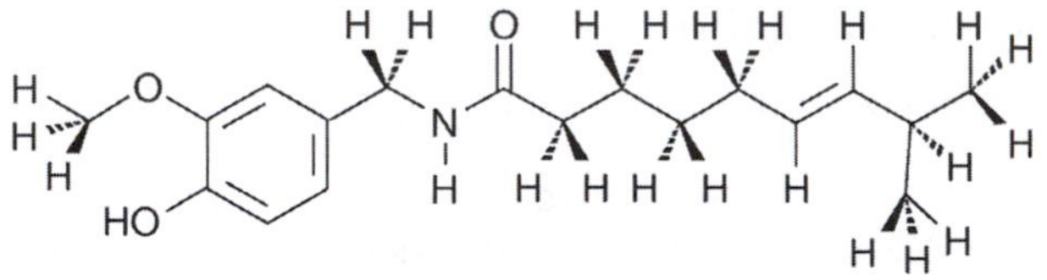

Figure   7.2.1 ▶   The molecular structure of capsaicin N-[4-hydroxy-3-methoxyphenyl)methyl]-8-methyl-6-noneamide.

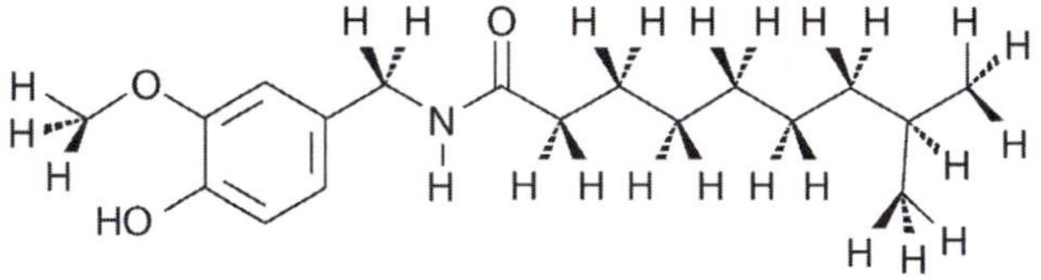

Figure 7.2.2 ▶   The molecular structure of dihydrocapsaicin.

homocapsaicin (capsaicin with a $(CH_2)_5$ unit instead of $(CH_2)_4$), and *homodihydrocapsaicin* (dihydrocapsaicin with $(CH_2)_7$ instead of $(CH_2)_6$). The capsaicinoid content varies with the pepper. In bell peppers capsaicin and dihydrocapsaicin are present in about a 1:1 ratio, while in Tabasco peppers the ratio is closer to 2:1.

KEY TERMS: solubility   hydrogen bonding

Related Web Sites

▶ The Chile Pepper Institute, Capsaicin Research Around the World,
   http://www.chilepepperinstitute.org/homepage.htm

7.3 Why Are Bubbles in a Carbonated Drink Spherical?

Imagine a stream of bubbles rapidly rising to the surface of glass of soda. What shape are these bubbles? Without any prompting, all of us would picture spherically shaped bubbles. What chemical principles preclude oval-shaped or cubic-shaped bubbles?

The Chemical Basics

Bubbles in a soft drink primarily arise from the carbonation, or dissolved carbon dioxide gas, added prior to bottling the soda. The gas bubbles are surrounded by liquid, primarily composed of water. The water dictates the shape of the gas bubble. From a energy standpoint, water molecules prefer to interact with other water molecules and not with the carbon dioxide gas molecules. The dissimilarities in chemical structure and properties of liquid water and gaseous carbon dioxide create an unfavorable interaction of these distinct substances. To reduce the amount of interaction of water with gas, a spherical bubble forms. Why? For a given volume of space, the shape that has the lowest surface area is a sphere. By inducing the gas to form a spherical bubble, the water surrounding the bubble minimizes the surface area of the gas bubble with which it must interact. In this way, fewer water molecules must surround the gas bubble, maximizing the number of water molecules that remain in close proximity to only other water molecules.

The Chemical Details

The spherical shape of bubbles arises from the sizable surface tension of the liquid soft drink (primarily water). The surface tension of a liquid is a measure of its resistance to increase its surface area. Liquids with relatively large intermolecular forces tend to have relatively high surface tensions, that is, high resistances to increasing their surface area. Why do the sizable intermolecular forces between water molecules necessarily dictate a spherical shape for the enclosed gas bubble? In a soft drink a water molecule at a distance from a bubble is able to experience intermolecular attractions to other water molecules surrounding it. On the other hand, a water molecule at the surface of a gas bubble has fewer neighboring liquid molecules with which it may undergo intermolecular attractions. As water molecules have strong attractive intermolecular forces (hydrogen bonding), the positioning of water molecules at the surface of a gas bubble requires some energy, since some intermolecular forces must be overcome. For a given volume a sphere has a smaller surface area than any other shape, thus the minimum energy cost to form a bubble occurs when the surrounding liquid molecules form a spherical "enclosure" about a gas bubble. Bubbles formed in liquids of surface tension lower than that of water may often be observed to have nonspherical shapes. Soap solutions, solutions of water with added surfactant that lowers the surface tension of the water, can form bubbles of distorted spherical shapes.

KEY TERMS: **hydrogen bonding**   **intermolecular forces**   **surface tension**

7.4 How Do Furniture Polishes Repel Dust?

"Dustblock formula actually repels dust." "Cleaning, dusting, and polishing are all completed in seconds." You've heard the claims of various furniture polishes. Is there actually a chemical basis to dusting?

The Chemical Basics

Furniture polishes are used to dust, clean, shine, and protect wood surfaces. The ingredients within the polish formulation confer these favorable attributes. In particular, to act as a dust repellent, the polish contains antistatic ingredients, that is, hydrocarbon substances that are not prone to static electricity and therefore do not attract dust. Silicone oil[1,2] and lemon oil are typical "antistatic" ingredients.

Dust particles carry an electrical charge and are attracted to any surface that develops a charge. There are two key ways for a surface to acquire a charge. Friction causes the accumulation of static electricity on surfaces, and thus a surface subjected to frictional forces is susceptible to dust build-up. Additionally, the direct carrying of electrical current by electronic equipment causes such devices to become charged and therefore likely to gather dust. Furniture items attract dust from the charge acquired by frictional forces, and thus the application of an antistatic polish to the furniture surface reduces the likelihood of static electricity. Have you ever wondered why furniture makers discourage "dry dusting," i.e., dusting without polish? Such action, while perhaps temporarily removing a layer of dust, can cause microscopic scratches and even enhance the level of static electricity on the surface.

The Chemical Details

What happens when static charge develops as a consequence of frictional forces between two surfaces? A charge separation occurs, with electrons migrating from one surface to the other. The electron-deficient surface becomes positively charged, while the surface gaining electrons acquires an overall negative charge. Hydrocarbons exhibit weak intermolecular forces — dispersion forces — because of their nonpolar nature and their low *polarizability*, i.e., ease with which an electric field can deform the electronic charge distribution. Antistatic ingredients such as the hydrocarbon-based silicone oil and lemon oil thus deter the creation of friction-induced electrical charges.

Other synonyms for silicone oil, a mixture of many ingredients, include silicone, siloxane, and polydimethylsiloxane. In actuality, siloxanes are a diverse class of industrial polymers with a backbone of alternating silicon and oxygen atoms. Organic substituents, commonly methyl ($-CH_3$) groups and phenyl ($-C_6H_5$) rings, are attached to the tetravalent silicon atoms. The repeating unit in the most common siloxane polymer, polydimethylsiloxane, is given in Fig. 7.4.1.

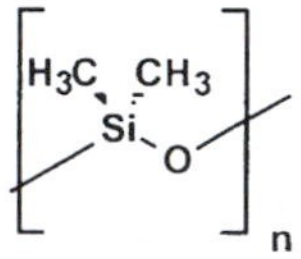

Figure 7.4.1 ► The repeating unit in the most common siloxane polymer, polydimethylsiloxane.

Lemon oil contains limonene or methyl-4-isopropenyl-1-cyclohexene, as shown in Fig. 7.4.2.

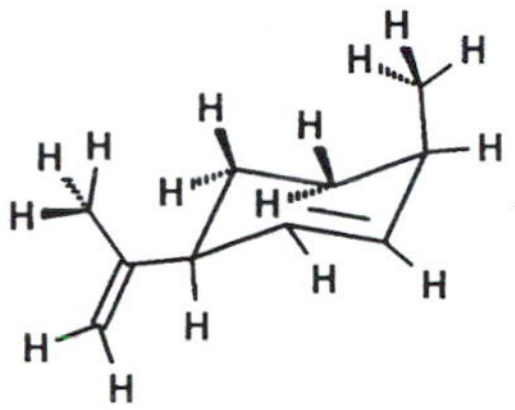

Figure 7.4.2 ► Limonene or methyl-4-isopropenyl-1-cyclohexene, a common ingredient of lemon oil.

KEY TERMS: **dispersion forces   nonpolar   polarizability
static electricity**

References

[1] "S C Johnson Wax — Pledge — Wax, Spray, Refinish, Aerosol, Material Safety Data Sheet."
http://www.biosci.ohio-state.edu/~jsmith/MSDS/PLEDGE.htm

[2] "Boyle-Midway — Old English Lemon Cream Furniture Polish, 1865, Material Safety Data Sheet." http://www.biosci.ohio-state.edu/~jsmith/MSDS/OLD%20ENGLISH%20LEMON%20CREAM%20FURNITURE%20POLISH.htm

Related Web Sites

► "Static Electricity." Electrostatics, Incorporated,
http://www.electrostatics.com/page2.html

► "Introduction to Static Electricity." http://www.simco.nl/start.htm

7.5   How Do Windshield Coatings Improve Visibility in a Rainstorm?

Aquapel Glass Treatment improves a driver's ability to see clearly and drive safer. This innovative technology is a long-lasting rain repellent.

—Aquapel Glass Treatment: Rain Repellent, Website copy

How can a product claim to improve your vision through an automobile windshield? The chemical structure of the key ingredients in the formulation can ensure a clear view during a rainstorm.

The Chemical Basics

A number of glass treatments are available for both automobiles and aircraft to disperse rain, sleet, and snow; prevent the build-up of frost, ice, salt, mud, bugs, oil, and road grime; and reduce glare. These glass coatings are durable but not permanent and can be reapplied. These rain-repellent materials contain *hydrophobic* substances that interact strongly with glass but readily shed rain drops.

The Chemical Details

The manufacturers of windshield coatings take advantage of the fact that the *hydrophilic* substances possess chemical structures that permit favorable intermolecular interactions with water. Chemical species capable of exhibiting hydrogen bonding, dipole–dipole interactions, or ion–dipole interactions with water are typically hydrophilic substances. Alternatively, *hydrophobic* substances typically are nonpolar molecules that exhibit only weak van der Waals interactions with water.

Two major classes of hydrophobic chemical substances can be applied to glass in ultrathin layers to inhibit surface wetting. *Siloxanes* or *polysiloxanes* or *silicones* are polymers with a "backbone" of alternating silicon and oxygen atoms. These macromolecules are quite chemically inert, show resistance to water, and exhibit stability at high and low temperatures. The most common siloxane polymer, polydimethylsiloxane, is composed of the monomeric (i.e., repeating) unit illustrated in Fig. 7.5.1.

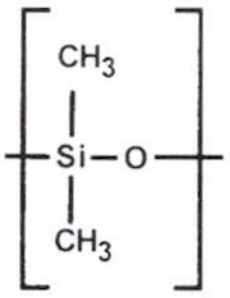

Figure 7.5.1 ▶ The monomeric (i.e., repeating) unit of the most common siloxane polymer, polydimethylsiloxane.

A hydroxy-terminated poly(dimethylsiloxane), $HO[-Si(CH_3)_2O-]_nH$, is one of the constituents of the formulation known as Permatex.[1]

The halogenated hydrocarbons known as *chlorofluorocarbons* are a second class of materials that comprise water-repellent coatings. In particular, the substance known as 1,1,2-trichloro-1,2,2-trifluoroethane or CFC-113 (Fig. 7.5.2) is

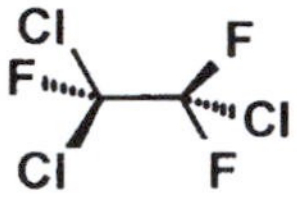

Figure 7.5.2 ▶ 1,1,2-trichloro-1,2,2-trifluoroethane or CFC-113.

the main ingredient of one windshield product.[2] Other proprietary fluorinated compounds are used in patented technology to produce a highly effective wind-shield glass treatment.[3]

KEY TERMS: hydrophobic   hydrophilic   siloxanes   chlorofluorocarbons

---

References

[1] Permatex Silicone Windshield & Glass Sealer Material Safety Data Sheet, Permatex Industrial Corporation, http://hazard.com/msds/mf/perm/files/81730.TXT

[2] "Potential Rainboe Toxicity in Commercial Aircraft Cockpits. Victoria Voge, MD, MPH, Gonzales, Tex; and Lance Schaeffer, JD, http://www.sma.org/smj/96aboccp.htm

[3] "Aquapel Glass Treatment News: Research Highlights Improved Safety Potential of Hydrophobic Windshield Coatings." PPG Industries, Inc., http://www.aquapel.com/

7.6 How Is a Fabric Made Water-Repellent or Waterproof?

Designing effective water-repellent, water-resistant, or waterproof fabrics to provide protection in inclement weather or during certain outdoor activities requires an understanding of the chemical and physical properties of water. Whether you are sailing in wind-driven rain, hiking in a downpour, or sitting on a wet surface, chemistry can keep you dry!

The Chemical Basics

A variety of finishes are applied to the surfaces of fabrics to impart a resistance to water. The water-repellent finish must be capable of interacting chemically with the fabric to form a film or coating, but the film must not possess a chemical structure favorable to interaction with water. Waxy coatings, silicone, and Scotchgard are some of the chemical finishes used in the fabric industry. In reality, over 100 different Scotchgard products with over 30 formulations have been designed to optimize the performance on a variety of fibers, including polyester, nylon, cotton, rayon, polypropylene, wool, and blends. The latest Scotchgard formulation — an environmentally friendly product — was introduced in 2001.[1] Du Pont has introduced a Teflon fabric protector to resist water stains on a variety of fabrics including silk. While most water-repellent finishes are surface treatments, some protective polymers are capable of penetrating the fabric and encapsulating every fiber throughout the fabric. Treatment with such polymers provides increased resistance to water.

Alternatively, recent developments in the textile industry have designed fabrics that repel water due to the structure of the fabric itself rather than a chemical finish applied to the fiber. Gore-Tex fabric is one example of a waterproof material consisting of a Teflon film (actually, a patented Gore-Tex membrane) glued or laminated to a conventional synthetic fabric such as nylon or polyester. The pores of the membrane are quite small — nine billion pores exist per square inch! In particular, each pore is 20,000 times smaller than a raindrop, but 700 times larger than a molecule of water.[2] As a consequence, a droplet of liquid water is too large to pass through a pore. The surface tension of water — a property of the liquid that arises from the extensive interactions between water molecules in the liquid state ("hydrogen bonding") — keeps the droplet intact, preventing tiny droplets from forming which could pass through the pores. On the other hand, water vapor from perspiration on the skin can pass through the pores since, as a gas, the water vapor molecules do not collect into droplets.

The Chemical Details

The application of water-repellent finishes to fabrics actually involves a chemical reaction between the material and the finish. Cellulose-based fibers such as cotton possess hydroxyl (–OH) groups that exist on the surface of fabrics spun and woven from the fiber. The basic structure of cellulose portrayed in Fig. 7.6.1 reveals

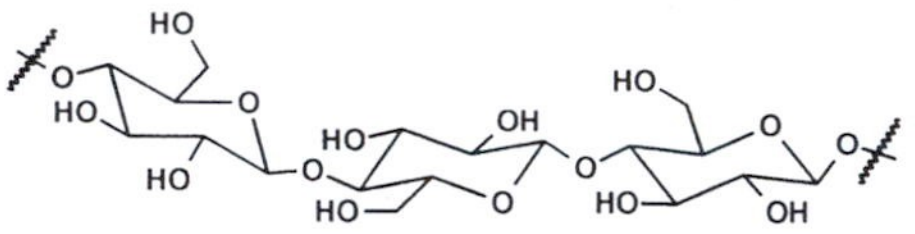

Figure 7.6.1 ▶ Representation of the polymer cellulose.

three hydroxyl groups per six-membered ring. As an example of a chemical reaction between cellulose and a water-repellent finish, consider the organosilicon compound trimethylchlorosilane, $(CH_3)_3SiCl$. This compound possesses strong Si–C bonds but in the vapor phase is capable of reacting with –OH groups to form oxygen–silicon linkages and generate HCl:

$$-OH + (CH_3)_3SiCl\ (g) \rightarrow -O-Si(CH_3)_3 + HCl\ (g).$$

Briefly exposing a cotton fabric to trimethylchlorosilane vapor replaces the strong water-attracting hydroxyl groups with water-repellent $-O-Si(CH_3)_3$ groups on the fabric surface. The loss of hydrogen bonding capacity is the prime factor for the water-resistant character of the treated fabric. Fluorocarbon-based finishes such as Scotchgard are applied to fibers in much the same fashion. The fluorochemical is composed of two major parts: a water-soluble portion that reacts with a functional group on the fiber (such as the hydroxyl group in cellulose) and an inert portion (due to the strength of carbon–fluorine bonds) that repels water.

Waterproof Gore-Tex fabric is an innovative application of chemistry that provides an extremely useful consumer material. The technology developed by W. L. Gore and Associates uses unique microstructures to control the porosity of the fabric. The patented Gore-Tex membrane is a composite of two materials — expanded polytetrafluoroethylene (ePTFE) to provide the membrane structure and polyalkylene oxide polyurethane-urea (POPU) to protect the membrane from soiling or contamination. Both the physical and chemical characteristics of the PTFE membrane give rise to the waterproof nature of Gore-Tex fabric. The pore structure of the membrane acts as an impermeable barrier to liquid water while the hydrophobic ("water-hating") nature of the polymeric PTFE repels water. The nonpolar carbon-fluorine bonds of the PTFE membrane do not permit hydrogen bonding between water and the fluorine atom. Liquid water molecules retain an affinity for each other rather than for the PTFE membrane, thereby maintaining water droplet size. As the membrane structure is integral to the functioning of Gore-Tex fabric, contamination or blockage of the pores by body oils, soaps, lotions, cosmetics, insect repellants, and dry cleaning solvents must be avoided. Saturation of the membrane with the oleophobic ("oil-hating") substance POPU ensures that contamination does not occur. Lamination of the Gore-Tex membrane to a conventional fabric adds durability.

KEY TERMS: hydrogen bonding   hydrophobic

References

[1] "3M Scotchgard Cleaner for Rugs & Carpet (Cat. No. 1019) MSDS Sheet." http://multimedia.mmm.com/mws/mediawebserver.dyn?lllll2liq 25lq_5jKDGwuuR0X&lFrm4uqmlhsZ

[2] "FAQS: Gore-Tex Products." The W. L. Gore & Associates, Inc., http://www.gorefabrics.com

Related Web Sites

▶ The W. L. Gore & Associates, Inc. Homepage, http://www.gorefabrics.com

▶ 3M Innovation Network — The Homepage of the 3M Company, http://www.mmm.com/profile/looking/index.html

▶ "Innovative Lives: Patsy Sherman and the Invention of Scotchgard." Smithsonian Institution, http://www.si.edu/lemelson/centerpieces/ilives/ lecture11.html

7.7   How Does a Bullet-Proof Vest Work? How Is It Made?

Security personnel often rely on the integrity of bullet-proof vests to protect their lives. The innovative chemistry of the fabric creates the incredible strength of these ballistic barriers.

The Chemical Basics

Bullet-resistant apparel are often constructed of manmade fibers known as *aramids*. Kevlar, developed by DuPont in 1971, is one example of this class of fibers. The unique properties exhibited by Kevlar—ultrahigh strength and stiffness—arise from its chemical composition. How can chemical identity impart such features as structural rigidity, cut resistance, flame resistance, even inertness to chemical attack? Kevlar is an example of a synthetic polymer composed of long, rod-like chains of stiff chemical units ("monomers") that are strongly linked ("bonded"). An extensive network of interactions between the linear chains imparts further rigidity and stability. This combination of intermolecular (i.e., between two or more molecules) interactions and intramolecular (i.e., within the same molecule) bonds gives Kevlar a strength per weight that is five times that of steel. In addition to ballistic body armor, there are numerous applications for the strength of fibers like Kevlar. As a durable yet lightweight and flexible fiber, Kevlar is used in the manufacture of gloves and sleeves that offer protection from heat and cuts. Protective clothing for firefighting and for sports such as bicycle helmets, batting gloves, and NHL goalies' face masks are additional apparel uses for Kevlar. Sporting equipment including canoes and kayaks, kites, golf clubs, ski poles, and in-line skates use Kevlar as a composite fiber to increase durability, add stiffness, and reduce weight. As a reinforcement fiber, Kevlar also offers high performance for puncture-resistant tires for demanding terrain and for wear-resistant automotive brake pads and hoses. Airships constructed from Kevlar are sealed on the inside with a film such as saran or mylar.

   On the horizon is a new technology for ballistic protective clothing using nonwoven fabrics. Since the weaving process can be detrimental to the linear alignment that gives fibers such as Kevlar their strength, the scientists at AlliedSignal have designed a process that maintains the fiber orientation. Both aramid fibers and polyethylene-based fibers such as Spectra fiber designed by the chemists at AlliedSignal may be employed in this technology. How is this nonwoven fabric created? Parallel strands of the linear synthetic fiber are mounted side-by-side with a flexible *thermoplastic resin* (i.e., a polymer that attains the characteristics of plastic upon heating). Two such layers of unidirectional fibers are arranged at 90° to cross the fibers at right angles and then fused with heat and pressure to create a composite structure. Two sheets of thin, flexible polyethylene film are then sandwiched on either side of the composite structure to reduce contact with dirt, moisture, and abrasives. These laminated rolls of fabric are ultralight and per weight ten times stronger than steel. Fabricated into durable composite structures, Spectra-based products are providing increased effectiveness as ballistic barriers.

The Chemical Details

Aramids are polymeric species that contain both aromatic and amide groups. The aromatic functionality consists of a substituted benzene ring, while the amide moiety is a C=O carbonyl group bonded to a nitrogen atom with at least one hydrogen substituent, C(=O)–NHR. The combination of the distinct chemical properties of these two functionalities provide the basis for the ultrahigh strength of the polymer. The aromatic portion of the polymer confers strength within a single polymer chain by limiting bond rotation, while the amide linkage is essential for strong hydrogen bonding between polymer chains. Kevlar, the trademark for poly-*para*-phenylene terephthalamide, is an example of an aramid polymer formed from two different monomeric units, a *diamine* (i.e., a monomer with two amine or –NH$_2$ groups) and a dicarboxylic acid (i.e., a monomer with two –COOH or carboxyl groups). The condensation reaction of terephthalic acid and 1,6-hexamethlenediamine produces the amide linkage and a water molecule as by-product, as shown in Fig. 7.7.1. Mixing a stoichiometric amount of diamine

Figure   7.7.1 ▶   The condensation reaction of terephthalic acid and 1,6-hexamethlenediamine to produce the amide linkage in Kevlar and a water molecule as by-product.

and dicarboxylic acid (i.e., a 1:1 ratio) yields a linear polymer with the repeating unit shown in Fig. 7.7.2. *Why is this linearity important to the functioning of the*

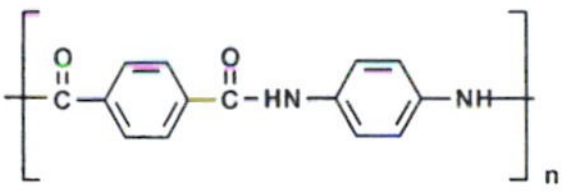

Figure 7.7.2 ▶   The repeating unit of the polymer Kevlar.

polymer? The rod-like nature of the polymer orients the molecules in solution in essentially a single direction, facilitating the strong interchain hydrogen bonding needed to create the cross-linked matrix (Fig. 7.7.3).

The first synthesis of Kevlar by solution polymerization was reported by S. L. Kwolek, P. W. Morgan, and W. R. Gorenson of DuPont in U. S. Patent 3,063,966 (1962). In 1980, Stephanie Kwolek won the American Chemical Society's Award for Creative Invention, and on July 22, 1995, she was inducted into the National Inventors Hall of Fame in Akron, Ohio. In 1996 Stephanie Kwolek was awarded

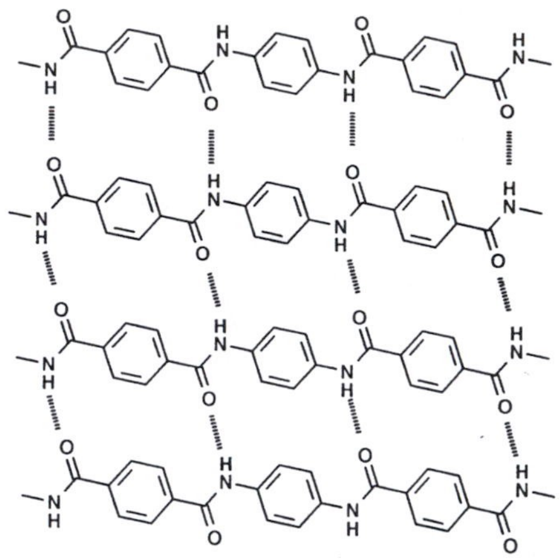

Figure 7.7.3 ▶   Strong interchain hydrogen bonding in Kevlar.

the National Medal of Technology "for her contributions to the discovery, development and liquid crystal processing of high-performance aramid fibers which provide new products worldwide to save lives and benefit humankind."[1]

KEY TERMS: **amide bond or amide linkage   hydrogen bond   aramid**
condensation polymer or condensation reaction

References

[1] "National Medal of Technology; 1985–1997 Recipients."
      http://www.ta.doc.gov/medal/Recipients.htm

Related Web Sites

▶ DuPont Kevlar Home Page, http://www.dupont.com/afs

▶ "Kevlar: The Wonder Material." Microworlds: Exploring the Structure of Materials, Lawrence Berkeley Laboratory, http://www.lbl.gov/MicroWorlds/Kevlar/

▶ "A Big Hand for Kevlar Plus." Du Pont On-Line Magazine, July 6, 1997, http://www.dupont.com/kevlar/europe/kevlar_WP/01_00/hand_c.html

▶ Bulletproof Vests, First Defense International, http://www.firstdefense.com/html/bulletproof_vests.htm

▶ "Innovative Lives: Stephanie Kwolek and the invention of Kevlar." Smithsonian Institution, http://www.si.edu/lemelson/centerpieces/ilives/ lecture05.html

▶ "Stephanie Louise Kwolek." Inventor of the Week Archives: The Lemelson-MIT Prize Program, http://web.mit.edu/invent/www/inventorsI-Q/kwolek.html

7.8 Why Is Cotton so Absorbent and Why Does It Dry So Slowly?

You have no doubt noticed many differences between natural fibers like cotton and synthetic fibers like polyester. Cotton is the most commonly used fiber by designers and manufacturers. U.S. cotton producers and importers of cotton goods into the United States have emphasized this point in their advertisements for cotton, "The Fabric of Our Lives."[1] Nevertheless, the desirable characteristics of polyester have made it the most used and most blended synthetic fiber. The chemical structures of these fibers provide an explanation for the differences in the response of these materials to water.

The Chemical Basics

The effectiveness of a towel in absorbing a water spill depends on the chemical structure of the fiber. One of the governing principles in chemistry is that, for any two substances, the closer their chemical structures, the greater the likelihood of the materials interacting. The attractive interactions on the molecular level are called intermolecular forces. Water is a highly polar substance with an affinity for other polar materials. Polarity is enhanced when the atoms, bonds, and shape of a molecule lead to an unequal distribution of charge. In addition, a water molecule has the ability to form strong associations with other water molecules through interactions known as "hydrogen bonds." Cotton fibers are also capable of linking with each other or with water through hydrogen bonds. These compatible associations lead to high water absorbency by cotton towels. In contrast, the chemical structure of synthetic polyester fibers does not favor as strong a connection on the molecular level with water, leading to reducing absorbency. The same principles explain why lightweight synthetic fabrics dry much faster than cotton towels and natural fibers.

The Chemical Details

The fiber cotton is composed of the polymer cellulose whose structure appears in Fig. 7.8.1. All of the factors required for *intermolecular* (i.e., between two molecules) *hydrogen bonding* are present in cellulose. In particular, recall that a hydrogen bond consists of an arrangement of three atoms, denoted X–H • • • Y, where the symbol • • • represents the hydrogen bond and X and H are atoms covalently bonded to one another. X is typically the element F, O, or N, that is, a small electronegative element to create a polar bond with the element H. For intermolecular hydrogen bonding the atom Y must be an electronegative element

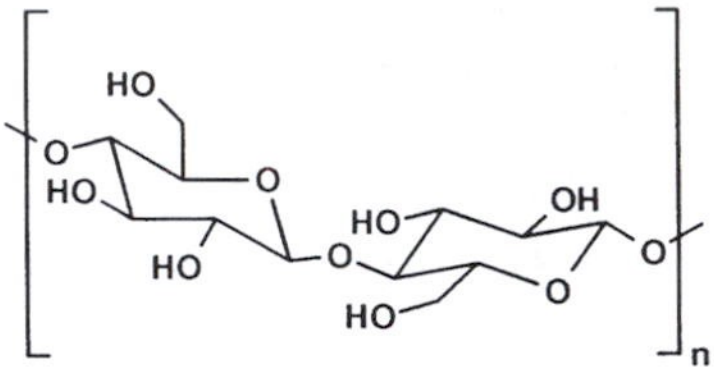

Figure 7.8.1 ▶ The repeating unit of the polymer cellulose, the constituent of the fiber cotton.

with an unshared pair of electrons and be contained within a second molecule either identical to or distinct from the molecule containing the X–H fragment. The strong polar nature of the X–H bond (due to the differences in electronegativities of X and H) results in a partial positive charge on the hydrogen. This partial charge separation may be denoted as $X^{\delta-}$–$H^{\delta+}$. The slight positive charge on H attracts the lone pair of electrons on Y, creating the so-called hydrogen bond. Colinearity of X, H, and Y yields the strongest possible hydrogen bond.

   The structure of cellulose reveals the two key components for hydrogen bonding: OH functional groups (the X–H unit) and additional oxygen atoms (the Y atom). Hence, water molecules may form intermolecular hydrogen bonds with cellulose in one of two ways:

(1) The central oxygen atom in water (acting as Y) can hydrogen bond the hydrogens in the O–H functional groups of cellulose (the X–H unit) or

(2) The hydrogen atoms in water (covalently bound to oxygen and thus composing an X–H fragment) can hydrogen bond to the oxygen atoms in cellulose (serving as Y).

   Thus, extensive intermolecular hydrogen bonding is possible between cellulose and water, enhancing the water capacity of a cotton towel.

   As an example of a polyester fiber, consider the condensation polymer Dacron (also sold as a film — Mylar). The monomeric repeating unit of Dacron is shown in Fig. 7.8.2, with several linked monomers indicated in Fig. 7.8.3. While the

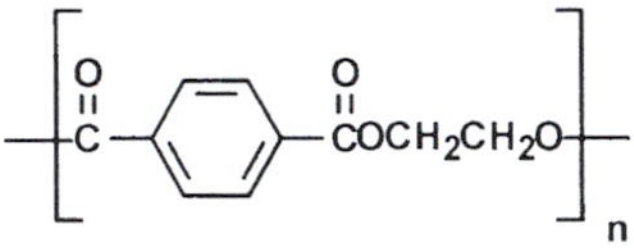

Figure 7.8.2 ▶ The repeating unit of the condensation polymer Dacron, an example of a polyester.

polyester fiber contains the electronegative oxygen atom (Y) to hydrogen bond to the hydrogen atoms in water (X–H), there are no corresponding X–H units in the polyester to hydrogen bond to the oxygen atom in water (Y). Hence, the extent of

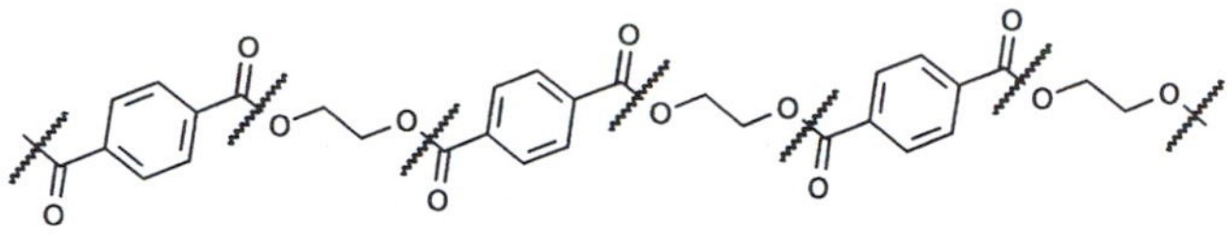

Figure 7.8.3 ►   Several linked monomers of the polymer Dacron.

hydrogen bonding between water and polyester is significantly reduced from the degree of hydrogen bonding possible between water and cellulose. This explains the relative effectiveness of the two materials in cleaning up water spills.

KEY TERMS: **intermolecular forces**   **hydrogen bonding**   **polymer**

References

[1] "Still the Fabric of Our Lives." J. Berrye Worsham, III, CEO/PresidentCotton Incorporated, Cotton Growers, February 2002, http://www.cottoninc.com/reprints/homepage.cfm? PAGE=3107

7.9   What Makes a No-Tears Shampoo?

Shampoo manufacturers provide a wealth of products to satisfy a range of consumer preferences. What kind of chemical ingredients are included to ensure a mild shampoo and guarantee a "no-tears" formula?

The Chemical Basics

As the central function of a shampoo is to cleanse the hair, the primary ingredient of a shampoo is a detergent (also known as a *surfactant*). Many shampoos, particularly those targeted for babies and children, claim to cause no eye irritation or sting. A "no-tears" formulation achieves this claim by carefully adjusting the nature of the surfactants. In particular, the identity and concentration of surfactants with an ionic or "charged" portion are controlled to minimize both eye and skin irritation.

The Chemical Details

A surfactant is a molecule that is characterized as amphiphilic, i.e., containing both a discrete hydrophilic (water-soluble) or polar portion and a well-defined hydrophobic (oil-soluble) or nonpolar fragment. The hydrophilic portion of the molecule is called the *surfactant headgroup*; the hydrophobic portion of a surfactant is described as the surfactant tail. The tail group is generally a linear long-chain hydrocarbon residue, such as the linear dodecyl group, $-C_{12}H_{25}$. Other hydrophobic groups are possible, including substituted benzene and other aromatic rings. The polar headgroup is generally classified as one of four types, depending on the charge on the hydrophilic group; anionic, cationic, nonionic, and amphoteric or zwitterionic headgroups are possible. Anionic surfactants carry a negatively charged headgroup and are extremely useful in shampoo because of their excellent cleansing, foaming, and water solubility properties. Anionic surfactants also rinse easily from the hair. These surfactants are also relatively inexpensive and easy to synthesize. For these reasons, anionic surfactants are the most common surfactants in personal care products. Nevertheless, anionic surfactants are known to be harsh and irritating to both eyes and scalp. Alkyl sulfates are the most frequently used anionic surfactants, including sodium (Fig. 7.9.1), ammo-

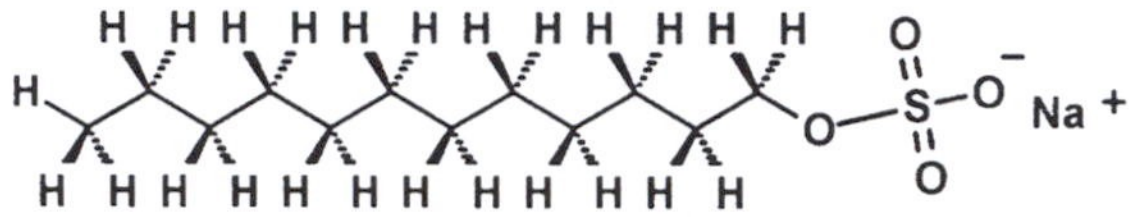

Figure 7.9.1 ▶ The molecular structure of the anionic surfactant sodium lauryl sulfate.

nium (Fig. 7.9.2), and triethanolammonium (TEA) (Fig. 7.9.3) lauryl sulfates.

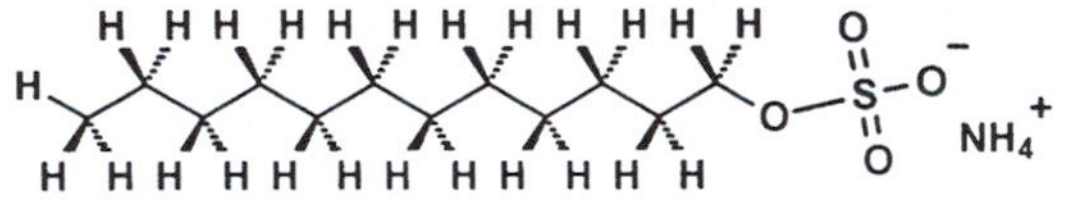

Figure 7.9.2 ▶ The molecular structure of the anionic surfactant ammonium lauryl sulfate.

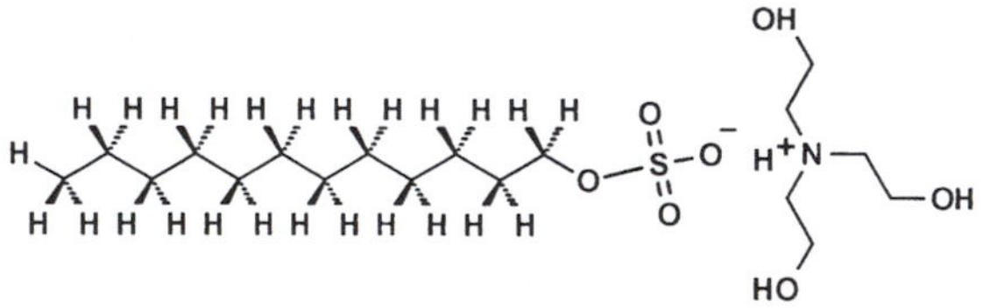

Figure 7.9.3 ▶ The molecular structure of the anionic surfactant triethanolammonium (TEA) lauryl sulfate.

Cationic surfactants contain a positively charged headgroup and are typically used as conditioners to improve hair manageability and reduce static. Cationic surfactants are especially irritating to eyes when used in high concentrations but are safe and useful in low amounts. Quaternium-15 (chloroallyl methanamine chloride, Fig. 7.9.4) is a cationic surfactant included in shampoo formulations

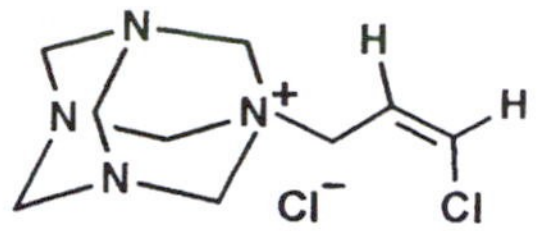

Figure 7.9.4 ► The cationic surfactant quaternium-15 (chloroallyl methanamine chloride).

for its conditioning ability. Nonionic (i.e., neutral) surfactants are not primarily used for their cleansing properties but to improve the solubility, foaming action, and conditioning action of a shampoo formulation. One nonionic detergent that is especially effective at reducing eye irritation is TWEEN 20 Polysorbate 20 (Fig. 7.9.5).[1] Amphoteric or zwitterionic surfactants (containing both positive

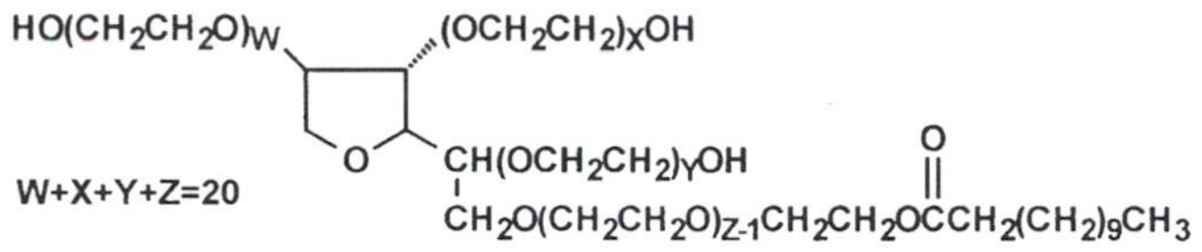

Figure 7.9.5 ► The anionic surfactant TWEEN 20 Polysorbate 20.

and negative charges) are used for their low foaming and slight irritation characteristics. Gentle "no-tears" shampoos contain significant amounts of amphoteric surfactants. An example of an amphoteric surfactant found in shampoo is lauramidopropyl betaine (Fig. 7.9.6). The net charge on the headgroup of a surfactant is

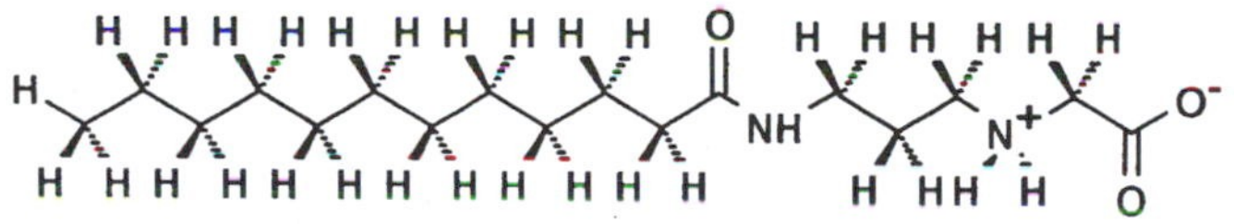

Figure 7.9.6 ► The amphoteric surfactant lauramidopropyl betaine.

one parameter that affects eye irritation. Experimental studies also suggest that a relationship exists between eye irritation and surfactant concentration.[2]

KEY TERMS: **surfactant**

References

[1] "Personal Care: Shampoos." Uniqema, http://surfactants.net/formulary/uniqema/pcm6.html

[2] G.Y. Patlewicz, R.A. Rodford, G. Ellis, and M.D. Barratt, "A QSAR model for the Eye Irritation of Cationic Surfactants." *Toxicology in Vitro* **14** (2000), 79–84.

Related Web Sites

- ▶ "Shampoos." The World of Hair, Dr. John Gray, P&G Hair Care Research Center, http://www.pg.com/science/haircare/hair_twh_116.htm

- ▶ "The Structure of Your Hair." http://www.geocities.com/HotSprings/4266/chem.html

7.10 How Does Milk Froth for Cappuccino Coffee?

Little Miss Muffet,
Sat on a tuffet,
Eating her curds and whey;
Along came a spider,
Who sat down beside her
And frightened Miss Muffet away.

—in *Mother Goose*, Kate Greenaway, 1881

While Little Miss Muffet may have enjoyed the whey proteins in milk in her own way, modern consumers enjoy whey proteins as the important element in the foam of a cappuccino.

The Chemical Basics

Coffee connoisseurs often enjoy a frothy cup of cappuccino, a thick, creamy concoction of coffee topped with a cap of velvety foam. Interestingly, the first use of the word "cappuccino" is traced to the Italians in the 16th century,[1] referring to the long, pointed cowl or cappuccino (derived from cappuccio meaning "hood") that was worn as part of the habit of the Capuchin order of friars founded by Saint Francis of Assisi. The restoration of Catholicism in Reformation Europe is attributed to the Capuchin monks whose order was established after 1525. In more modern usage in the mid-to-late 1940s, the word cappuccino was introduced to describe espresso coffee mixed or topped with steamed milk or cream, so called because the color of the coffee resembled the color of the habit of a Capuchin monk.[1]

Cappuccino lovers would argue that the foam cap on the drink is the critical element to a great cappuccino. Steam frothing of milk to prepare a cappuccino coffee involves injecting air and steam into milk to create the foam and to heat the milk to near boiling.

The foaming capacity of milk is related to the ability to form stable air bubbles. One factor that can stabilize air bubbles is the presence of a film coating on the bubble. Whey proteins, primarily β-lactoglobulin, can form stabilizing films on air bubbles. In their normal conformation, these proteins are globular; upon heating, however, intermolecular forces are disrupted to denature the proteins, that is, unfold the proteins. In the unfolded state whey proteins are able to form a thin film that is the surface of the air bubble.[2,3] The normal surface tension of water is so high that water forms droplets rather than films on surfaces. The presence of the protein film lowers the surface tension of water, that is, reduces the cohesive forces between water molecules, allowing water to spread and form films which are the walls of bubbles. The protein film coating also gives resistance to shearing of the bubble by providing a higher level of stretching capacity. Thus, the steaming of milk facilitates frothing, i.e., the process of making bubbles. Steam frothing can be further improved by using homogenized milk.[4] Homogenization is a process that splits fat globules into particles no larger than 2 μm.[2]

What are some factors that can reduce the foaming capacity of milk? One such factor is the degradation of milkfat, either naturally or via the presence of contaminating bacteria.[4] The breakdown of milkfat produces free fatty acids known as monoglycerides and diglycerides. These free fatty acids act as surfactants ("surface-acting agents") and associate with the whey protein, reducing the film-coating capacity of whey proteins on air bubble surfaces and thereby reducing the frothing capacity of milk.

The Chemical Details

The principal milk proteins are caseins (contributing about 80% of the total milk proteins) and whey proteins. Interestingly, whey proteins are susceptible to denaturation from an increase in temperature; casein proteins are not heat denatured. Casein proteins have very little tertiary structure, that is, a defined three-dimensional spatial arrangement. These proteins have a significant number of "exposed" hydrophobic amino acids, as observed in an unfolded (denatured) protein (hence the reason for the heat tolerance). Whey proteins are characterized by a compact, globular conformation and a molecular weight that ranges from 14,000 to 1,000,000 g per mole.[2] Two common whey proteins are β-lactoglobulin and α-lactalbumin. β-Lactoglobulin contains 162 amino acids with a molecular weight of about 18,300 g per mole. This protein exhibits a pH-dependent structure.[5] A monomeric form predominates at pH values below 3.0 and above 8.0. An association of monomers via carboxyl groups to form an octamer occurs between pH 3.1 and 5.1. At other pH values, including the pH of milk (about 6.5–6.7),[6] a spherical dimer with a diameter of approximately 18 Å predominates. The whey protein α-lactalbumin consists of 123 amino acids and has a molecular weight of 14,146 g per mole.[4] Although denatured by heat, α-lactalbumin exhibits resistance to denaturation in the presence of calcium. The protein uses calcium to form intramolecular ionic bonds that tend to make the molecule resistant to thermal unfolding. Other whey proteins include serum albumin with a molecular weight of about 69,000 g per mole, and immunoglobulins with two amino acid chains with molecular weights of 20,000–25,000 and two amino acid chains with molecular weights of 50,000–70,000.

KEY TERMS: intermolecular forces   proteins   colloidal dispersion

---

References

[1] "Did You Know That?" koffeekorner.com,
http://www.koffeekorner.com/didyouknow.htm

[2] "Nutrition and Food Management 236: Milk." Oregon State University,
http://osu.orst.edu/instruct/nfm236/milk/index.cfm#b

[3] "Nutrition and Food Management 236: Lecture: Whey Protein Concentrates." Oregon State University, http://class.fst.ohio-state.edu/FST822/lectures/WPC.htm

[4] Background Briefing: Steam Frothing of Milk, Dairy Chemistry, The Institute of Food Science & Technology, Dairy Research and Information Center, University of California — Davis, http://drinc.ucdavis.edu/html/dairyc/index.shtml

[5] "Whey Proteins." Ohio State University, Food Science 822,
http://class.fst.ohio-state.edu/FST822/lectures/Milk2.htm

[6] "Dairy Chemistry and Physics." Dairy Science and Technology, University of Guelph,
http://www.foodsci.uoguelph.ca/dairyedu/chem.html#acid

7.11   Why Is Sickle Cell Anemia a "Molecular Disease"?

The Nobel Prize chemist Linus Pauling related the mechanism of sickle cell anemia to a genetic defect in hemoglobin synthesis and thus defined the first molecular disease. Pauling's groundbreaking paper in 1949 was boldly titled "Sickle Cell Anemia: A Molecular Disease."[1] Many claim that this discovery laid the foundation for molecular biology. What is the chemistry that is at the heart of this pioneering work?

The Chemical Basics

Sickle cell anemia is a disease that restricts blood flow to organs of the body and thereby prevents red blood cells from delivering necessary oxygen. Usual symptoms include swelling in the joints, severe pains in the legs, and jaundice. Linus Pauling called sickle cell anemia a molecular disease because the disease is due, in effect, to a molecular defect. Persons affected with this malady have an alteration in one single amino acid in a long chain of amino acids in the protein known as hemoglobin. This single change causes significant effects in the ability of this protein to carry oxygen in the blood.

The Chemical Details

Normal hemoglobin molecules are complex, three-dimensional structures consisting of four chains of amino acids known as polypeptide chains. Two of these chains are known as alpha subunits with 141 amino acid residues each, and the remaining polypeptide chains are the beta subunits with 146 amino acid residues each. The sequences of amino acids in the alpha and beta subunits are different, but fold up via noncovalent interactions to form similar three-dimensional structures. When a polypeptide chain arranges itself in space, i.e., "when it folds," amino acids that were far apart in the chain are brought closer in proximity. The final overall shape of the protein molecule is influenced by (1) the amino acids in the chain, and (2) the interactions that are possible between distant amino acids.

One of the amino acids in the beta polypeptide chain of hemoglobin is glutamic acid (Fig. 7.11.1).

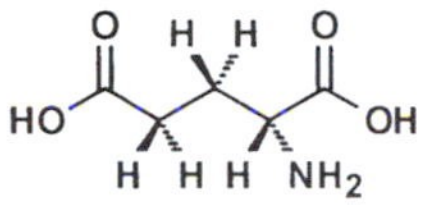

Figure 7.11.1 ▶ The molecular structure of glutamic acid.

The hydrophilic polar side group on glutamic acid is capable of forming hydrogen bonds with water and thus helps to keep normal hemoglobin dispersed within the red blood cells.

In sickle cell hemoglobin, the glutamic acid of the beta subunit is replaced by the amino acid valine (Fig. 7.11.2). Even though only this one amino acid is

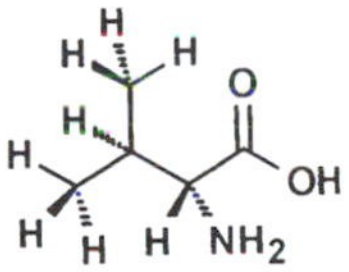

Figure 7.11.2 ▶ The molecular structure of valine.

altered, the change has profound effects.

How does the side group of valine differ from that of glutamic acid? The valine side group, an isopropyl group, is a nonpolar hydrophobic group that cannot form hydrogen bonds with water. Located on the outside surface of the folded protein, this nonpolar valine interacts favorably with other nonpolar residues (particularly phenylalanine and leucine) in neighboring hemoglobin protein molecules.[2] In essence, this one amino acid change causes the protein to repel water and be attracted to other protein molecules. Mutated valine residues in other sickle cell hemoglobin molecules also seek to stabilize their structures by interacting with other hydrophobic residues. These aggregations gradually lead to the formation of long protein filaments that distort the shape of red blood cells. The red blood cells, normally disk-shaped, take on a crescent or sickle shape. These deformed cells clog narrow capillaries, thereby restricting blood flow (and hence oxygen delivery) to organs of the body, leading to the disease sickle cell anemia.

KEY TERMS: **intermolecular forces**   **amino acids**   **hydrogen bonds**
hydrophobic

References

[1] Pauling L., Itano H. "Sickle cell anemia, a molecular disease." *Science* **110** (1949), 543–548.

[2] "Sickle Cell Anemia Tutorial." University of Illinois, http://peptide.ncsa.uiuc.edu/tutorials_current/Sickle_Cell_Anemia/

Related Web Sites

▶ "The Hemoglobin Protein: A Useful Teaching Model." Department of Biochemistry, Medical College of Wisconsin, http://www.biochem.mcw.edu/science_ed/Pages/hemoglobin/real1.html#useful

Other Questions to Consider

| 3.8 | How do fog machines create the artificial fog or smoke used in theatrical productions? *See* p. 23.

| 5.7 | What puts the "blue" in "blue jeans"? *See* p. 45.

| 13.17 | How can sucralose, an artificial sweetener made from sugar, contain *no* calories? *See* p. 205.

| 14.6 | What is the difference between hard and soft contact lenses? *See* p. 221.

Chapter 8 ▶

Connections to Phases and Phase Transitions

What Do Meteorologists Use to "Seed" Clouds?

Rainmaking: while the term invokes images of primitive rituals to influence the rain gods, modern day meteorologists have sound chemical principles to induce precipitation. What chemistry fundamentals are instrumental in cloud seeding?

The Chemical Basics

Clouds form as masses of warm, moist air rise by convection into cooler regions, causing the water vapor to condense into liquid water droplets or ice crystals. Condensation normally occurs on microscopic particles suspended in the air, such as sea-salt particles, clay-silicate particles from land, and smoke particles produced by combustion processes such as forest fires. These particles are known as *cloud condensation nuclei*. What causes the release of precipitation from clouds? The water droplets or ice crystals must grow to sufficient size to fall from the cloud at a speed to avoid evaporation during their fall through the air to reach the ground. Many clouds composed of liquid droplets limited to a few tens of micrometers are stable for long periods of time. The collision of cloud droplets and the aggregation process that occurs to form larger particles is known as *coalescence*. In clouds with ice crystals, these particles serve as condensation nuclei. Weather modification is often attempted by meteorologists to combat extreme situations such as drought. Measures such as cloud seeding are often effective when supercooled clouds exist, that is, clouds composed of liquid water droplets at temperatures below the freezing point of water. Atmospheric conditions routinely result in clouds at temperatures as low as -10 or $-20°C$, well below the normal freezing point of water. If ice crystal formation occurred, as expected at the given temperature, the crystals would fall out of the clouds and melt at higher temperatures closer to earth. To induce precipitation in supercooled clouds, substances are introduced to act as condensation nuclei around which water droplets coalesce.

In 1946 Irving Langmuir and Vincent J. Schaefer, chemists working at the General Electric Research Laboratories in Schenectady, N.Y., discovered that dry ice pellets could induce the formation of ice crystals in a cloud composed of water droplets in a deep-freeze box. The ice crystals continued to grow in size and eventually dropped to the bottom of the container. Grains of dry ice can be dispersed from airplanes to create the same effect in the atmosphere. Other substances, notably silver iodide, mimic the structure of ice and can serve as a nucleus for ice crystal formation. These substances can be introduced into clouds via aircraft or alternatively from the ground using rockets carrying a pyrotechnic substance ingrained with silver iodide. In addition, when these iodide salts are burned in air, a smoke of tiny particles is generated that can be carried upward by air currents. It could be supposed that wilderness wildfires are often extinguished by rainfall that is induced through the seeding of clouds by smoke particulates. Seeding with silver iodide particles has also been used to mitigate or suppress the formation of hailstones in cumulonimbus clouds. By increasing the number of hail particles and thereby decreasing their size, the destructive effects are reduced. Silver iodide has also been effective at dispersing fog at airports, inducing the formation of ice crystals and the precipitation of snow. Results have generally been inconclusive, however, for seeding programs to reduce the intensity of hurricanes.

The Chemical Details

Some of the most effective substances that have been found for cloud seeding are solid carbon dioxide and silver iodide. As solid CO_2 at $-77°C$ drops through a supercooled cloud layer, the water droplets in the cloud instantaneously freeze. The ice crystals continue to grow in size as water vapor condenses on the solid particles, eventually reaching the dimensions necessary to fall as precipitation. Why is solid silver iodide so effective at nucleating ice crystals in supercooled clouds? The answer lies in the similarity of the crystal lattice structure of silver iodide to that of ice. While at least nine structurally different forms of ice are known, "normal" hexagonal ice predominates at the temperatures and pressures where clouds form. In this structural form of ice, each oxygen atom is surrounded by a nearly tetrahedral arrangement of four other oxygen atoms in neighboring water molecules. One of the crystal forms of silver iodide, AgI, has a hexagonal ZnO or wurtzite structure similar to that of ice (Fig. 8.1.1).[1] Thus, water can form ice crystals growing on the AgI seed crystals, just as if the water were growing on existing ice crystals.

KEY TERMS: **supercooled liquid**   **crystal structure**   **silver iodide**

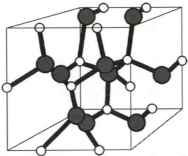

Figure 8.1.1 ▶ The ZnO or wurtzite crystal structure illustrating the tetrahedral arrangement of oxygen atoms about each zinc atom.

References

[1] N. N. Greenwood and A. Earnshaw, "Copper, Silver, and Gold," in *Chemistry of the Elements*, 2nd ed. (New York, Pergamon Press: 1997), Chap. 28, p. 1173.

Related Web Sites

- "The Physical Basis for Seeding Clouds." Atmospherics, Inc.,
  http://www.atmos-inc.com/weamod.html

- "Irving Langmuir." Woodrow Wilson Leadership Program in Chemistry, The Woodrow Wilson National Fellowship Foundation,
  http://www.woodrow.org/teachers/chemistry/institutes/1992/Langmuir.html

- The discovery of cloud seeding, United States Bureau of Reclamation Environmental Education, http://www.rsgis.do.usbr.gov/html/history.htm

- "Does Weather Modification Really Work?" Ric Jensen, Information Specialist, Texas Water Resources Institute, http://twri.tamu.edu/twripubs/WtrResrc/v20n2/text.html

---

8.2 | # Why Do Citrus Growers Spray Their Trees with Water to Protect Them from a Freeze?

During freeze warnings in certain agricultural areas, we often hear that farmers spray their crops with water to provide protection from the cold temperatures. What chemical principles are growers applying to protect their crops?

The Chemical Basics

The warm-weather climates necessary for citrus orchards generally minimize the risk of frost conditions. Nevertheless, occasionally measures are needed to protect the trees and their fruit from freezing temperatures. Citrus crops become threatened when temperatures fall below 28°F for four hours or more.[1] Heating is obviously the most effective protection against frost, and often heaters are used to warm the air temperature. Microjets under trees also spray warm water onto the tree trunks to keep trees warm. As the warm water cools, heat is released to warm the air surrounding the crop. The hotter the water, the smaller the volume of water needed to provide the same degree of protection. In fact, a hot water system at temperatures near 150°F may be able to raise the orchard temperature by as much as 6°F under some conditions.[2] However, the phase change of liquid water to solid ice also generates heat. Thus, by spraying trees with water under conditions where freezing occurs will produce ice on the leaves and release enough heat to protect the tree and its fruit. The insulating layer of ice on plants and fruit also protects the crop by keeping the fruit temperature at 32°F.

The Chemical Details

The liquid → solid phase transition known as freezing is characterized by a negative enthalpy change (called the enthalpy of fusion), $\Delta H_{\text{freezing}} < 0$. At constant pressure, the enthalpy change of the freezing process is exactly equal in magnitude to the heat released as the phase transition occurs. (In other words, $\Delta H_{\text{freezing}} = q_p$ = heat transferred at constant pressure.) The exothermic (i.e., releasing heat) nature of the freezing process generates sufficient heat to protect the tree and its fruit:

$$H_2O \text{ (l)} \rightarrow H_2O \text{ (s)} \qquad \Delta H_{\text{freezing}} = -6.01 \text{ kJ/mol}.$$

Thus, one mole of water releases 6.01 kJ of heat as it freezes[3]; each gram of water releases 334 J upon freezing.

The use of warm water (i.e., water at temperatures above the freezing point) also generates heat as the water cools to its freezing point. The exact amount of heat released depends on the amount of water present and its temperature according to the formula

$$\Delta H_{\text{cooling}} = C_{\text{p, water (l)}} \Delta T = C_{\text{p, water (l)}} (T_{\text{liquid water}} - T_{298 \text{ K}}).$$

The heat capacity of liquid water at 20°C, for example, is 4.2 J g^{-1}K^{-1}.[4] Thus, for every degree above the freezing point of water, one gram of water releases 4.2 J upon cooling one degree.

KEY TERMS: phase transition   heat capacity   enthalpy of fusion

References

[1] Mark Wilcox, "Preventing Cold Injury." Yuma County, Cooperative Extension, University of Arizona,
http://ag.arizona.edu/aes/citrusnews/Orchard%20Establishment%20articles/Orchard%201.htm#ColdInjury

[2] Geraldine Warner, "Hot Water Could Protect Trees from Cold Damage." *Good Fruit Grower*, February 15, 1997.

[3] "Enthalpy of Fusion," in *Handbook of Chemistry and Physics*, 74th ed., ed. David R. Lide (Boca Raton, FL: CRC Press, 1993), **6**–116, **6**–125.

[4] "Properties of Water in the Range 0–100°C," in *Handbook of Chemistry and Physics*, 74th ed., ed. David R. Lide (Boca Raton, FL: CRC Press, 1993), **6**–10.

Related Web Sites

► "Orchard Smudge Pots Cooked Up Pall of Smog." South Coast Air Quality Management District, http://www.aqmd.gov/monthly/smudge.html

8.3 What Is the Dark Spot Left on the Inside of a Light Bulb When it Burns out?

We are striking it big in the electric light, better than my vivid imagination first conceived. Where this thing is going to stop Lord only knows.

—Thomas Edison, October 1879

In his lifetime, Thomas Alva Edison patented 1,093 inventions, but the incandescent light bulb is generally regarded as his most famous invention. On October 21, 1879, 29-year-old Thomas Edison demonstrated the first incandescent lamp in Menlo Park, NJ. The bulb burned for 13.5 hours. The chemistry of the components of this lighting device are essential to its incredible success.

The Chemical Basics

The dark spot on the inside of a burned-out light bulb is tungsten metal that has sublimed (vaporized from the solid) as a consequence of the heating of the tungsten filament to produce white light.

The Chemical Details

Incandescent lamps consist of glass bulbs that enclose an electrically heated filament that emits light. For over 50 years prior to Thomas Edison's success, scientists had experimented with developing electric lamps. With financiers such as J. P. Morgan and the Vanderbilts, Edison founded the Edison Electric Light Company in 1878 with the prime mission to generate cheap electric power to provide an illumination source. For the filament of his electric lamp, Edison reportedly[1] experimented on 6000 different types of materials, eventually narrowing his focus on fine platinum wire and a mix of 10% iridium with platinum. Unfortunately, these filament choices were unsuccessful, for the materials would not handle the current without melting. Edison's final (and successful) incandescent lamp utilized carbonized cotton filaments. Modern incandescent lamps employ tungsten filaments. What properties of a material enhance its use as a filament in incandescent light bulbs? Materials with a high resistance to the flow of electricity and a high melting temperature are ideal, for the resistance will cause heat to be generated in the material until it glows white. A high melting temperature ensures that the glow is maintained. Because hot metals emit only small amounts of light in the infrared range, a metal filament maximizes the amount of visible light emitted and minimizes infrared radiation that would generate heat.[2]

What physical and chemical properties of tungsten make this metal an ideal choice for the filament material? As a practical consideration, the ductility of tungsten enables the production of filaments that consist of fine wires. Furthermore, since an electric current generates high temperatures, a substance like tungsten with a high melting point permits a longer lifetime of incandescence. In fact, tungsten has the highest melting point of any element — 3683 K or 3410°C or 6170°F.[2] As light bulbs are often filled with a gas (argon and nitrogen, for example) to carry heat away from the filament, the filament material should be chemically inert with respect to these gases. Tungsten meets this criterion, exhibiting no chemical reactivity with these gases. By keeping the tungsten filament "cooler" with the presence of inert gas, less tungsten is lost by sublimation. Otherwise, the thinning of the filament by the sublimation loss of tungsten would increase the resistance to current flow, thereby increasing the filament temperature. Eventually, the resistance would reach a high enough level that the current required to produce light would essentially vaporize the tungsten metal. As an additional benefit, the increased pressure in the system due to the added gas also aids in reducing the extent of sublimation of the filament. "Long-life" incandescent light bulbs use different combinations of gases to fill the bulb in order to lengthen the life of the tungsten filament.

In contrast to normal lamps filled with inert gases, halogen lights are filled with gaseous iodine or bromine to take advantage of the chemical reactivity of tungsten and gaseous halogens. At the high temperatures (> 3000°C) near the tungsten filament in an operating halogen lamp,[3] tungsten does not combine chemically with these halogens. However, nearer the wall of the bulb where temperatures may be 800–1000°C cooler,[3] a gaseous tungsten halide forms. This vapor eventually migrates to the filament and decomposes at high filament temperatures, depositing tungsten metal on the glowing filament and regenerating the halogen gas. This ongoing cycle of chemical reaction and chemical decomposition to maintain the tungsten filament yields a brighter and longer-lasting light.

KEY TERMS: **incandescence**   **sublimation**

References

[1] "Adventures in Cybersound, Edison's Incandescent Electric Lamp, 1879." Dr. Russell Naughton, http://www.acmi.net.au/AIC/EDISON_BULB.html

[2] "Tungsten." Chemicool Periodic Table, David D. Hsu, http://wild-turkey.mit.edu/Chemicool/elements/tungsten.html

[3] "Scientific American: Working Knowledge: Halogen Lamps." Terry McGowan, July 1996, http://www.sciam.com/0796issue/0796working.html

Related Web Sites

▶ "The Undiscovered World of Thomas Edison." Kathleen McAuliffe, The Atlantic Monthly, **276**, (6) (December 1995), 80–93, http://www.theAtlantic.com/atlantic/issues/95dec/edison/edison.htm

▶ "Thomas Edison — Inventor." Edison-Ford Winter Estates, http://www.edison-ford-estate.com/thomas-edison/Ebio.htm

▶ "Thomas A. Edison Papers." Rutgers, The State University of New Jersey, http://edison.rutgers.edu/

▶ Antique Light Bulb Time Line: 1879 to 1925." Tim Tromp, http://www.bulbcollector.com/timeline.html

8.4   What are the Red or Silver Liquids in Thermometers?

We are accustomed to the rise and fall of the liquid in a thermometer as temperatures increase and decrease, respectively. Why do these liquids respond to temperature in this fashion, and what are the identities of these materials?

The Chemical Basics

The invention of the thermometer is generally credited to Galileo. His instrument, built near the end of the sixteenth century, relied on the expansion of air with an increase of heat. Traditional liquid-in-glass thermometers were devised in the 1630s and are standard equipment today in research settings, medical practice, and meteorological measurement.

Many common thermometers contain a liquid confined within a narrow capillary tube. The liquid height varies with the surrounding temperature. In actuality, the volume of the liquid is responding to temperature, and the liquid tries to expand equally in all directions. By confining the liquid in a tube, the only direction for ready expansion is along the length of the narrow tube. Thus, expansion in that direction (i.e., liquid height) can be used as a measure of the ambient temperature. Most liquids expand in volume as their temperature increases, and, because the extent of expansion is generally constant over a range of temperatures, the amount of expansion can be quantified and calibrated. In particular, two liquids exhibit a consistent and measurable expansion at commonly measured temperatures — liquid mercury and ethanol (also known as ethyl alcohol). Daniel Gabriel Fahrenheit (1686–1736), a German physicist and maker of scientific instruments, is credited for inventing the alcohol thermometer in 1709 and the mercury thermometer in 1714 (as well as developing the temperature scale that bears his name). These liquids permit common temperatures to be readily measured, such as the boiling and freezing of water. Why are these particular measurements possible? Mercury has a higher boiling point than water, and ethanol has a lower freezing point than water. The silvery color of mercury facilitates viewing the liquid level in a thermometer, but the colorless appearance of ethanol generally is modified with a red dye to enhance distinguishing the liquid level.

The Chemical Details

The useful temperature ranges for mercury and ethanol-filled thermometers depend on the temperature ranges over which these materials remain as liquids. Mercury exhibits freezing and boiling points at atmospheric pressure (i.e., normal freezing and boiling points at 1 atm) of -38.9 and $356.6°C$. Thus, a mercury thermometer is advantageous when high-temperature conditions are likely. Ethanol, with normal freezing and boiling points of -114.1 and $78.3°C$, respectively, is convenient as a thermometer liquid when temperatures below the freezing point of water are to be measured. The equation relating the volume change of a material to a change in temperature is given by

$$\Delta V = \beta V_o(\Delta T) \quad \text{or} \quad V_f = V_o(1 + \beta \, \Delta T)$$

for V_o at an initial T of $0°C$ and where β is the volume expansion coefficient or the coefficient of cubical expansion or the *coefficient of thermal expansion*. The units of β are reciprocal temperature, for β is defined as the increase in volume per unit volume per degree celsius rise in temperature. Mercury has an average value for of $1.8169041 \times 10^{-4} °C^{-1}$ over the temperature range of 0 to $100°C$ and $1.81163 \times 10^{-4} °C^{-1}$ over the temperature range of 24–299°C.[1] The volume expansion coefficient β for ethanol averages $1.04139 \times 10^{-3} °C^{-1}$ over the temperature range of 0–80°C,[1] a larger value allowing for finer calibrated thermometers. An important point to note is that any expansion or contraction of the thermometer container itself (usually glass) is generally ignored when calibrating household thermometers because liquids generally have a substantially larger coefficient of thermal expansion than do solids. As an example, Corning 790 glass exhibits a cubical expansion coefficient of $2.4 \times 10^{-6} °C^{-1}$.[2] Pyrex glass contains borosilicate glass, a type of glass that is exceptionally resistant to heat, expanding only about one-third as much as common silicate glass. As a consequence, Pyrex is often used to make chemical apparatuses, including thermometers.

KEY TERMS: normal boiling point   volume expansion coefficient

                normal freezing point   (or coefficient of cubical expansion)

References

[1] "Coefficients of Cubical Expansion for Various Liquids and Aqueous Solutions," in *Lange's Handbook of Chemistry*, 12th ed., ed. John A. Dean (New York, McGraw-Hill: 1979), Table 10-42.

[2] "Coefficients of Cubical Expansion of Solids," in *Lange's Handbook of Chemistry*, 12th ed., ed. John A. Dean (New York, McGraw-Hill: 1979), Table 10-43.

Related Web Sites

▶ "The Thermometer." Albert Van Helden,
http://es.rice.edu/ES/humsoc/Galileo/Things/ thermometer.html

8.5   What Nineteenth-Century "Disease" Destroyed Cathedral Organ Pipes?

Restorations of nineteenth-century organs in the cathedrals of northern Europe revealed a metal "disease" often attributed to the corrosion of tin. Chemically speaking, however, the structural change in the metal pipes is a completely different phenomenon. What aspects of chemistry must organ builders consider when attempting to achieve a particular acoustical character?

The Chemical Basics

Organ pipes today are made of a variety of woods (e.g., mahogany) and metal alloys, depending on the desired tonal quality, appearance, and cost. Most pipes are made of a varying mixture of tin and lead called *spotted metal*, but pipes with copper and zinc are also common.[1] It is widely believed that the higher the tin content, the brighter the tone and the shinier the appearance. Historically, tin was the preferred material for an organ pipe, yet the expense of the metal was always a consideration and a limitation.

Pure tin exhibits two common forms in the solid state — a *gray tin* and a *white tin*. At temperatures above 13°C or 55°F, the more stable form of tin is the denser white tin. At lower temperatures, the white tin is slowly converted to the gray form, a more powdery substance. Prolonged exposure to the cold winter temperatures of northern Europe contributed to the loss of integrity and disintegration of many cathedral organ pipes. As a consequence of the progressive nature of the structural transformation, as the white tin metallic surface becomes covered with gray powder, the degradation is often known as "tin disease," "tin pest," or "tin plague."

The Chemical Details

At atmospheric pressure, pure solid tin adopts two structures or allotropes, depending on temperature. At room temperature white metallic tin is stable but, at temperatures below 13°C, white tin undergoes a phase transformation into gray tin. White tin (also known as β-tin) adopts a body-centered tetragonal crystal structure (Fig. 8.5.1). Allotropic gray tin (α-tin) crystallizes in a cubic diamond

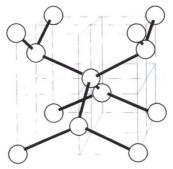

Figure 8.5.1 ▶ A body-centered tetragonal crystal structure adopted by white tin.

crystal structure (Fig. 8.5.2).

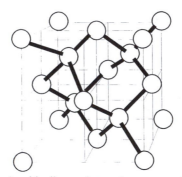

Figure 8.5.2 ▶ A cubic diamond crystal structure adopted by gray tin.

KEY TERMS: **allotrope**   **phase change**   **phase diagram**

References

[1] "Front Pipes, Principals, Mixtures & Mutations, Flutes and Strings." F. J. Rodgers Ltd., Craftsman Organ Pipe Makers and Voicers, http://www.musiclink.co.uk/pipeorgan/flues.html

Related Web Sites

► "Materials." David Harrison, School of Science & Technology, Athrofa Addysg Uwch Gogledd Ddwyrain Cymru, North East Wales Institute of Higher Education, Wrexham, North Wales, http://www.newi.ac.uk/buckleyc/materials.htm

► "The Science Corner: Allotropes." Nigel Bunce and Jim Hunt, College of Physical Science, University of Guelph, http://www.physics.uoguelph.ca/summer/scor/articles/scor40.htm

► "Behind The Pipes." D. A. Flentrop, http://www.chapel.duke.edu/organs/Flentrop/memorial/flentrop.htm

Other Questions to Consider

3.6   Why are ice cubes cloudy on the inside? *See* p. 20.

Chapter 9 ▶

Connections to Equilibria

9.1 What Causes the Fizz When an Antacid Is Added to Water?

The familiar fizzing action that occurs as an antacid tablet dissolves in water is the result of a chemical reaction involving the ingredients in the tablet.

The Chemical Basics

The effervescence of an antacid tablet in water is key to the effectiveness of the antacid. Two familiar antacids, Alka-Seltzer and Bromo Seltzer, contain sodium bicarbonate or baking soda — $NaHCO_3$. The bicarbonate ion reacts with citric acid, another ingredient of the product, to produce carbonic acid, H_2CO_3. In solution, carbonic acid decomposes to yield water and gaseous carbon dioxide (CO_2), the same gas in carbonated beverages and champagne. The fizz generated when the tablet dissolves (see color Fig. 9.1.1) is simply the CO_2 bubbles coming out of solution.

The Chemical Details

Each Alka-Seltzer tablet contains 1916 mg of sodium bicarbonate, 1000 mg of citric acid (Fig. 9.1.2), and 325 mg of aspirin or *acetylsalicylic acid* (Fig. 9.1.3).[1] Bromo Seltzer also contains sodium bicarbonate and citric acid as well as acetaminophen (Fig. 9.1.4). Several equilibria describe the action of bicarbonate-based antacid tablets. First of all, sodium bicarbonate dissolves completely in aqueous solution to generate sodium ions and bicarbonate ions:

$$NaHCO_3 \text{ (s)} \rightarrow Na^+ \text{ (aq)} + HCO_3^- \text{ (aq)}.$$

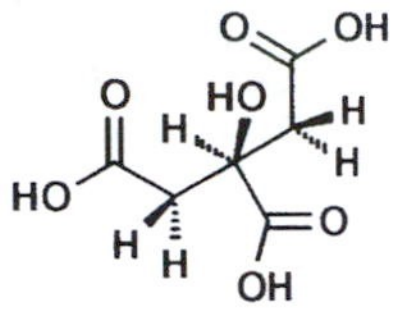

Figure 9.1.2 ▶   The molecular structure of citric acid.

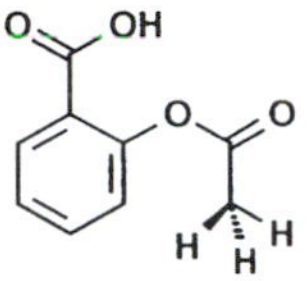

Figure 9.1.3 ▶   The molecular structure of aspirin or acetylsalicylic acid.

In water the bicarbonate ion is in equilibrium with carbonic acid:

$$HCO_3^- \text{ (aq)} + H_2O \text{ (l)} \quad \rightleftharpoons \quad H_2CO_3 \text{ (aq)} + OH^- \text{ (aq)}$$

$$K_{eq} \quad = \quad Ka(HCO_3^-) = 5.6 \times 10^{-11}.$$

The small value of the equilibrium constant indicates that the formation of carbonic acid is not very extensive in neutral water. However, the formation of carbonic acid is quite favored in acidic solution (arising from the citric acid also contained in the product):

$$HCO_3^- \text{ (aq)} + H_3O^+ \text{ (l)} \quad \rightleftharpoons \quad H_2CO_3 \text{ (aq)} + H_2O \text{ (l)}$$

$$K_{eq} \quad = \quad \frac{1}{K_a}(H_2CO_3) = \frac{1}{4.3 \times 10^{-7}} = 2.3 \times 10^6.$$

The carbonic acid produced via the reaction with stomach acid decomposes to yield dissolved gaseous CO_2:

$$H_2CO_3 \text{ (aq)} \rightarrow H_2O \text{ (l)} + CO_2 \text{ (g)}.$$

Production of H_2CO_3 by the action of bicarbonate ions on the acid in the stomach drives the above equilibrium to the right to generate CO_2 bubbles.

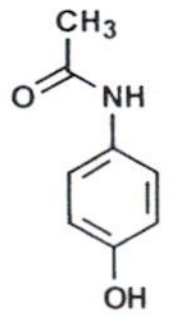

Figure 9.1.4 ▶   The molecular structure of acetaminophen.

> **KEY TERMS:**   equilibrium   equilibrium constant

References

[1] "Alka-Seltzer Original: Back of Package Information." Bayer Corporation,
    http://www.alka-seltzer.com/as/images/ASoriginal.gif

Related Web Sites

▶ "Alka-Seltzer: Student Science Experiments." Bayer Corporation,
   http://www.alka-seltzer.com/as/experiment/student_experiment.htm

Other Questions to Consider

| 3.4 | How is the toxicity of barium sulfate controlled for X-rays? *See* p. 17.

| 3.5 | Why is milk of magnesia an antacid? *See* p. 18.

| 9.2 | What is an antidote for cyanide poisoning? *See* below.

METAL COMPLEX IONS

9.2   What Is an Antidote for Cyanide Poisoning?

The high toxicity of cyanide-containing substances requires the immediate application of a medical antidote. What simple chemical
principles dictate both the toxicity of this species and the efficacy of
suitable antidotes?

The Chemical Basics

Cyanide (actually cyanide ion) is a particularly toxic substance as a consequence
of its strong affinity for the metal-containing enzymes responsible for providing
energy for cell respiration. In particular, cyanide forms stable complexes with the
metal iron (i.e., with ions of iron — ferric ions, Fe^{3+}) in the respiratory enzyme
cytochrome oxidase located in the mitochondria of cells. When the ferric ions are
complexed with cyanide, the functioning of cytochrome oxidase as an essential
catalyst for oxygen utilization in cells is inhibited. Normal cell functions cease
when cell respiration is impaired, leading to cell mortality. Antidotes for cyanide
poisoning offer cyanide ion the possibility of forming even more stable complexes
with a different metal ion, leaving the ferric ions free to function properly within
cytochrome oxidase.

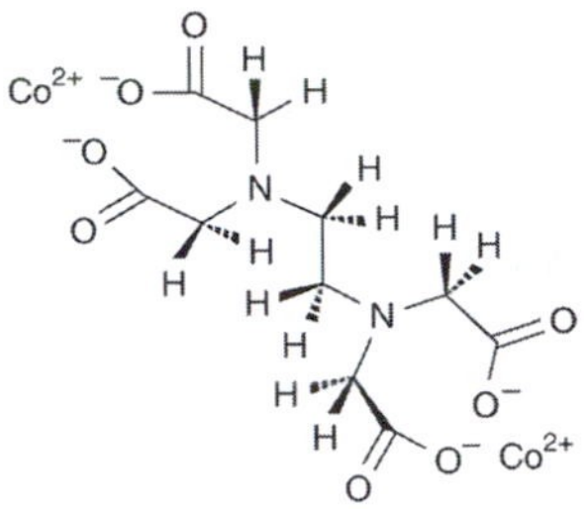

Figure 9.2.1 ▶ The molecular structure of dicobalt EDTA.

The Chemical Details

One antidote for acute cyanide poisoning used by toxicologists and physicians is the product known as Kelocyanor. One ampule of Kelocyanor contains 300 mg of the cobalt salt dicobalt edetate or dicobalt EDTA, Co_2EDTA (Fig. 9.2.1).[1] EDTA (ethylenediaminetetraacetic acid) is a multidentate ion with up to six lone pairs of electrons that are capable of forming six *coordinate covalent bonds* with metal ions. Four of these coordinating sites involve lone pairs of electrons on an oxygen atom in the acetate ligand, and two coordinating sites involve lone pairs of electrons on each of the nitrogen atoms. In dicobalt edetate, a neutral substance, each Co^{2+} ion is coordinated to two acetate ligands. The formation constant for this complex is large:[2]

$$2\,Co^{2+} + 4\,EDTA \;\rightarrow\; Co_2EDTA$$
$$K_{formation} \;=\; 2.0 \times 10^{16}.$$

The basis for the toxicological activity of this substance is the reaction of cobalt ion with cyanide ion to form a relatively nontoxic and stable ion complex. The hexacyanocobaltate ion contains a Co^{2+} central metal ion with six cyanide ions as ligands. This *coordination complex* involves six *coordinate covalent bonds* whereby each cyanide ion supplies a pair of electrons to form each covalent bond with the central cobalt ion. The formation constant for the hexacyanocobaltate ion is even larger than for dicobalt EDTA,[3] and thus the cobalt ion preferentially exchanges an EDTA ligand for six cyano ligands:

$$Co^{2+}\,(aq) + 6CN^-\,(aq) \;\rightarrow\; Co(CN)_6^{4-}\,(aq)$$
$$K_{formation} \;=\; 1.2 \times 10^{19}.$$

The exchange of EDTA for the CN^- ion reduces the concentration of cyanide ion in the body, making the cobalt ion an effective scavenger of toxic cyanide ions.

KEY TERMS:	complex formation	complex ion
	coordinate covalent bond	coordination complex

References

[1] "Antidotes: Kelocyanor." S.E.R.B. Laboratories, Paris, http://www.serb-labo.com/b3_b4_antidote.html

[2] D. A. Skoog, D. M. West, and F. J. Holler, *Analytical Chemistry: An Introduction*, sixth ed. (Philadelphia: Saunders College, 1994), 242.

[3] D. G. Peters, J. M. Hayes, and G. M. Hieftje, *Chemical Separations and Measurements* (Philadelphia: W. B. Saunders, 1974), A.12.

Related Web Sites

▶ "Cyanide Poisoning." S.E.R.B. Laboratories, Paris, http://www.serb-labo.com/c1_cyanide_poison.html

9.3 Why Is EDTA Added to Salad Dressings?

An inspection of the ingredients in many sandwich spreads, mayonnaises, margarines, and salad dressings reveals the abbreviation "EDTA." Even "real mayonnaise" has this important ingredient. The chemical structure of this substance helps this additive perform its important function as a preservative.

The Chemical Basics

The presence of unwanted metal ions in foods and beverages can often be traced to their presence in soils and in the machinery used for harvesting and processing of food. In particular, contamination by even trace amounts of copper, iron, or nickel is especially undesirable because these metals are known to catalyze the reaction of oxygen with unsaturated fats in foods, leading to undesirable color changes and rancidity. Food spoilage can be retarded or eliminated by including certain food additives known as *sequestrants*. These additives, where EDTA is an example, form tightly bound complexes with the trace metals, preventing the catalytic function of the metal ions and diminishing the degradation of the food item. Salad dressings, mayonnaise, margarine, processed fruits and vegetables, canned shellfish, and soft drinks are common food items with added EDTA.

The Chemical Details

The sodium and calcium salts of EDTA (ethylenediaminetetraacetic acid, Fig. 9.3.1.) are common sequestrants in food products. A three-dimensional representation of EDTA is shown in color Fig. 9.3.2. The EDTA ion is an especially effective sequestrant, forming up to six *coordinate covalent bonds* with a metal ion. These bonds are so named because a lone pair of electrons on a single atom serves as the source of the shared electrons in the bond between the metal ion and EDTA. The two nitrogen atoms in the amino groups and the oxygen

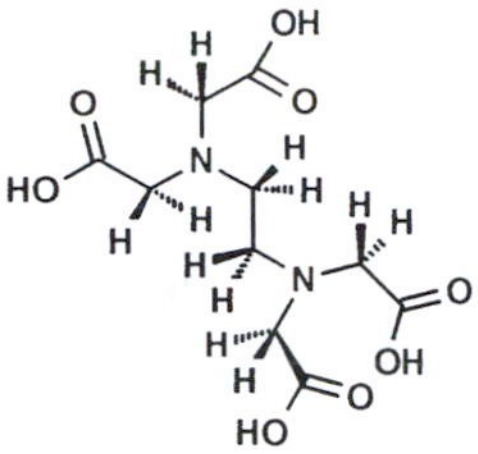

Figure 9.3.1 ► The molecular structure of ethylenediaminetetraacetic acid (EDTA).

TABLE 9.1 ► **Formation Constants for Complexes of EDTA with Metal Cations at 25°C**

Cation	K	Cation	K	Cation	K
Ag^+	2.1×10^7	Fe^{2+}	2.1×10^{14}	Mn^{2+}	6.2×10^{13}
Al^{3+}	1.3×10^{16}	Fe^{3+}	1.3×10^{25}	Ni^{2+}	4.2×10^{18}
Ca^{2+}	5.0×10^{10}	Hg^{2+}	6.3×10^{21}	Pb^{2+}	1.1×10^{18}
Cu^{2+}	6.3×10^{18}	Mg^{2+}	4.9×10^8	Zn^{2+}	3.2×10^{16}

atoms in the carboxyl groups of EDTA are the electron donors to create the coordinate covalent bonds that sequester the metal ion. One-to-one stoichiometry of metal–EDTA complexes leads to four, five, or six coordination positions around the central metal ion being occupied. Some values for the formation constants (K) of EDTA complexes involving metal cations are summarized in Table 9.1.[1]

The formation reaction is given by the equation

$$M + \text{EDTA} \rightleftarrows M \cdot \text{EDTA}$$

The larger the formation constant, K, the more likely the $M \cdot \text{EDTA}$ complex is to form and the fewer free metal ions remain.

A common form of EDTA used as a preservative is calcium disodium EDTA ($CaNa_2EDTA$). What metals will this form of the sequestrant scavenge effectively? The dissolution of the solid will yield calcium ions, sodium ions, and the EDTA anion. Any metal more effectively complexed than calcium will be readily scavenged, including all ions listed in Table 9.1 except silver (Ag^+) and magnesium (Mg^{2+}). (In the absence of the calcium counterion, as in the case of the acid form of EDTA, chelation of calcium in the body can occur. In fact, EDTA administered orally is an FDA-approved treatment for calcium deposits in the bloodstream that lead to cardiovascular disease.) Citric acid (Fig. 9.3.3) is another sequestrant of metal ions in foodstuffs.

KEY TERMS: **chelating agent** **sequestrant** **coordinate covalent bond**

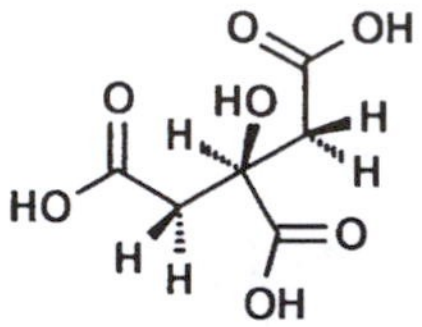

Figure 9.3.3 ▶   The molecular structure of citric acid.

References

[1] "Formation Constants for EDTA Complexes," in *Fundamentals of Analytical Chemistry*, 5th ed., ed. D. A. Skoog, D. M. West, and F. J. Holler (New York: Saunders College, 1988), 261.

Related Web Sites

▶ "Chemical of the Week: Chelating Agents." Department of Chemistry, University of Wisconsin–Madison, http://scifun.chem.wisc.edu/chemweek/ChelatingAgents/ChelatingAgents.html

▶ "Chelation Therapy: The History of EDTA." Leon Chaitow, N.D., D.O., http://www.healthy.net/library/books/chaitow/chelther/intro/history.htm

9.4   Why Do Hydrangeas Vary in Color When Grown in Dry versus Wet Regions?

Hydrangeas are popular summer flowers as a consequence of their vivid color that persists throughout the season. Large blooms of pink and blue colors are common, but the predominant color varies with a number of environmental factors. What role does chemistry play in determining the hue of these magnificent flowers?

The Chemical Basics

Certain varieties of hydrangeas are extremely popular for their ability to vary in color from blue to pink with growing conditions (see color Fig. 9.4.1). Research has determined that the actual mechanism for color variation is determined by the concentration of aluminum compounds in the flowers.[1] In the presence of aluminum, blue flowers result; in the absence of aluminum compounds, pink flowers predominate.

Many gardening books will state that the acidity of the soil affects the color of certain varieties of hydrangeas. The soil pH affects the availability of aluminum in the soil and thereby indirectly affects the flower color. A low pH or acidic conditions will yield blue blooms; pink blossoms will be favored by a higher pH or alkaline conditions. A purple color is the result of a more moderate pH level. Potting soils with a high level of peat moss will produce blue hydrangeas. Areas with significant rainfalls also produce blue flowers, probably due to the acidic nature of the rain water. In arid regions a pink color can result from the more alkaline content of the soil.

The Chemical Details

Soil acidity affects the availability of nutrients to the plant's root system. The solubility of a nutrient is often extremely sensitive to pH. What kinds of factors affect the pH of the soil? Some of the most important variables are the extent of rainfall, the presence of organic matter and certain microorganisms, the application of fertilizers, and the texture of the soil. The ideal pH range to promote blue hydrangeas is 5 to 5.5, while a higher pH of 6 to 6.5 is suited for pink blooms. For plants to develop blue flowers, aluminum ion (Al^{3+}) must be free to permit complexation with a pigment molecule to produce the blue color. Low pH ensures the solubility of aluminum ion. Under higher pH conditions, available aluminum ions in the soil will precipitate with hydroxide ions to form aluminum hydroxide and prevent metal complexation with the plant pigment, resulting in a pink color.

The plant pigments responsible for the variable coloration in hydrangeas is the class of pigments known as the red and blue *anthocyanins*. The anthocyanins are water soluble and display a color dependent on the acidity of their environment in the flower petal vacuoles. In acidic solution, the general structure of an anthocyanin is given by Fig. 9.4.2, with a formal charge on one oxygen atom and

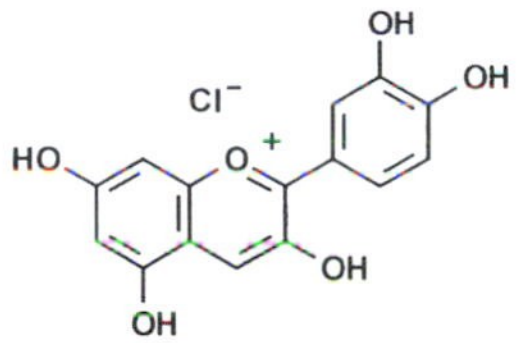

Figure 9.4.2 ▶ General structure of an anthocyanin pigment molecule in hydrangeas.

two neighboring (i.e., *ortho*) hydroxyl groups on one of the rings.[2]

Under these conditions, blue flowers can be achieved as a metal complex of anthocyanin (called a *metalloanthocyanin*) is stabilized through the association of a metal ion with two hydroxy groups oriented *ortho* to one another on the anthocyanin ring, as illustrated in Fig. 9.4.3.[2] In basic solution, the structure of the anthocyanin (Fig. 9.4.4) no longer has *ortho*-oriented hydroxy groups. Metal complexation is no longer possible, and the flower color appears red.

KEY TERMS: acid   base   metal complexation

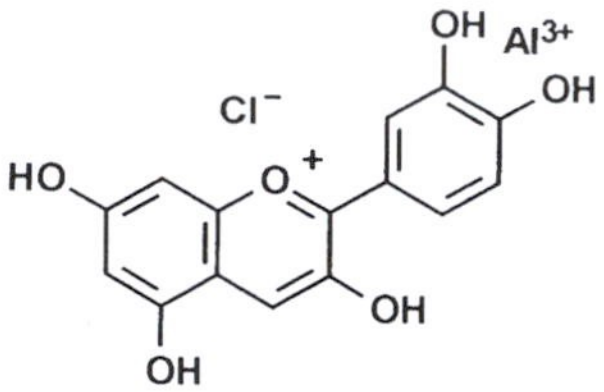

Figure 9.4.3 ▶ A metalloanthocyanin, a metal complex of anthocyanin stabilized through the association of a metal ion with two hydroxy groups oriented *ortho* to one another on the anthocyanin ring.

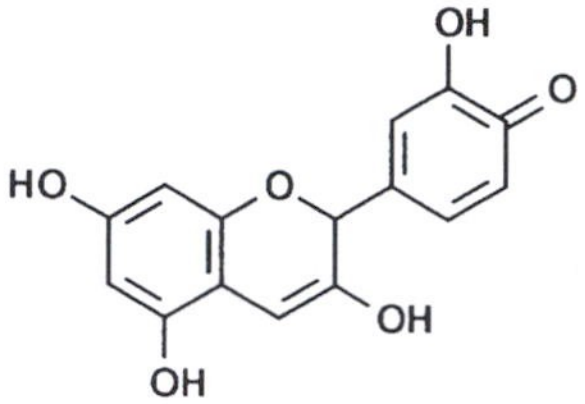

Figure 9.4.4 ▶ The structure of the anthocyanin pigment in basic solution, no longer possessing the *ortho*-oriented hydroxy groups.

References

[1] "Growing Bigleaf Hydrangea." Gary L. Wade, Extension Horticulturist, University of Georgia College of Agricultureal and Environmental Sciences, Cooperative Extension Service, http://www.ces.uga.edu/Agriculture/horticulture/hydrangea.html

[2] "The molecular basis of indicator color changes." F. A. Senese, Department of Chemistry, Frostburg State University, http://antoine.fsu.umd.edu/chem/senese/101/features/water2wine.shtml#anthocyanin

Related Web Sites

▶ "Soil pH and Fertilizers." Mississippi State Extension Service, http://msucares.com/pubs/is372.htm

▶ "Hydrangea Hydrangea macrophylla." Michigan State University Extension, http://www.msue.msu.edu/msue/imp/mod10/10000171.html

9.5 | Why Are Floor Waxes Removable with Ammonia Cleansers?

A high-quality metal cross-linked polymer is fortified for long-lasting durability.[1] A metal cross-linked floor finish is made from space-age polymer and designed to provide a highly durable floor finish with excellent gloss retention.[2] How has chemistry improved the modern floor finish?

The Chemical Basics

Modern floor finishes contain mixtures of ingredients that when applied as a liquid dry to give a clear, durable film coating. Most floor finishes are combinations of polymers blended to provide unique and desirable characteristics such as durability, gloss retention, resistance to scuffing, fast-drying, and ease of removal. Polymers made up of more than one monomer or repeating unit are called *copolymers*.

Polymeric surface coatings often gain their strength and durability from a three-dimensional network structure created by a cross-linking of the polymer chains. It is this network structure that can give floor waxes their unique properties. New formulations use metal ions to create the cross-linking structure. These metal ions are often soluble in ammonia cleansers, destroying the polymeric structure and allowing the floor wax to be "stripped."

The Chemical Details

The most common polymer groupings in floor waxes are acrylics, acrylic copolymers, and urethanes. An *acrylic polymer* is a generic term denoting derivatives of acrylic acid (CH_2=CHCOOH) and methacrylic acid (CH_2=C(CH_3)COOH), including acrylic esters (CH_2=CHCO$_2$R), and compounds containing nitrile (–CN) and amide (–CONH$_2$) groups such as acrylonitrile (CH_2=CHCN) and acrylamide (CH_2=CHCONH$_2$). A polyurethane is a polymer formed from linear repetitions of the monomer urethane ($CH_3CH_3OCONH_2$). One recent development is the "metal interlock floor finish," a fast-drying, durable, yet easily removed finish.[3] The floor wax formulations contain zinc,[4] often in the form of the *transitional metal complex ion* $Zn(NH_3)_4^{2+}$. As the liquid dries, the zinc ion is released upon evaporation of ammonia:

$$Zn(NH_3)_4^{2+} \text{ (aq)} \rightarrow Zn^{2+} \text{ (aq)} + 4\,NH_3 \text{ (g)}.$$

The zinc ion cross-links with the polymer to create sufficient strength and cross-linking density for durability and resistance to abrasion and detergents. The floor finish is easily removed with ammonia cleansers to reform the stable complex with zinc. The action of pulling the zinc out of the polymer allows the polymer to dissolve in the stripping solution.

KEY TERMS: transition metal complex ion   copolymers

cross-linked polymers

References

[1] "Floor Finishes." Zim International, http://chemicals.zim-intl.com/chemicals/
      floorfinishes.htm

[2] Floor Care Products, South Carolina Department of Corrections, Division of Industries,
      http://www.scprisonindustries.com/detail.asp?ID=158

[3] Floor Care Terminology, Clean Source, http://www.cleansource.com/floor.html

[4] "So What's the Real Problem with Zinc?" Tom Wright, Technical Director of Betco Cor-
      poration, 1991, http://www.michco.com/HelpStuff/Zinc_Problem_Help.html

Related Web Sites

▶ "Update: Floor Care Chemicals." Glen Franklin, Maintenance Solutions, May 1997,
   http://www.cleanlink.com/NR/NR2m7ea.html

▶ Case Study: Making a Case for Floor Care Technology, Steve Hanke,
   http://www.commercialfloorcare.com/CDA/ArticleInformation/features/BNP
   __Features__Item/0,3440,28274,00.html

9.6 How Do Drano and Liquid-Plumr Unclog Drains?

Drains run slower and slower as greases, soaps, fats, and detergents build up on the inner walls of drain pipes and eventually cause blockage. What is the chemical action of unclogging drains with household products such as Drano and Liquid-Plumr?

The Chemical Basics

Most household drain cleaners are caustic substances that create *alkaline* (basic) solutions when dissolved in water. Certain organic compounds commonly called grease dissolve in water that is highly basic. Thus, the alkalinity (i.e., characterized by a pH value above 7) of drain cleaners is responsible for the dissolution of grease that has clogged drains and plumbing. Some crystal drain cleaners also contain solid aluminum particles that react with an alkaline solution to produce hydrogen gas and release heat. Both the agitation of evolving gas and the release of heat during the chemical reaction produce additional forces to open up drains.

The Chemical Details

Commercially available drain cleaners such as Drano and Liquid-Plumr all contain sodium hydroxide (NaOH) as an active ingredient for dissolving grease.[1−5] Crystal drain cleaners contain the solid form of sodium hydroxide, while liquid drain cleaners are strong solutions of dissolved sodium hydroxide. In addition,

Figure 4.3.1 ▶ The Hope Diamond is the world's largest blue diamond and is on display in the National Museum of Natural History at the Smithsonian Institution. Photo reprinted with permission.

Figure 5.2.1 ▶ The Statue of Liberty prior to the restoration completed in 1986 (on the statue's 100th anniversary). The corroded copper outer surface is clearly visible. From Masterton and Hurley, *Chemistry: Principles and Reactions*, 4th edition. Orlando: Harcourt, 2001. Photo courtesy of Andy Levin/Photo Researchers, Inc.

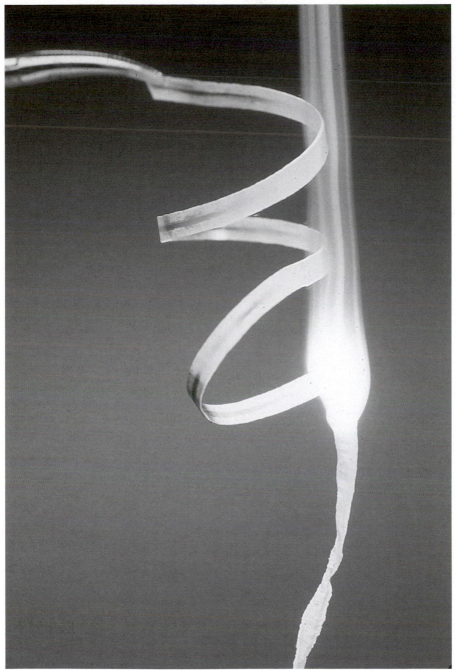

Figure 5.5.1 ▶ Formation of a white magnesium oxide coating as magnesium ribbon reacts with oxygen. From Masterton and Hurley, *Chemistry: Principles and Reactions*, 4th edition. Orlando: Harcourt, 2001. Photo courtesy of Charles D. Winters.

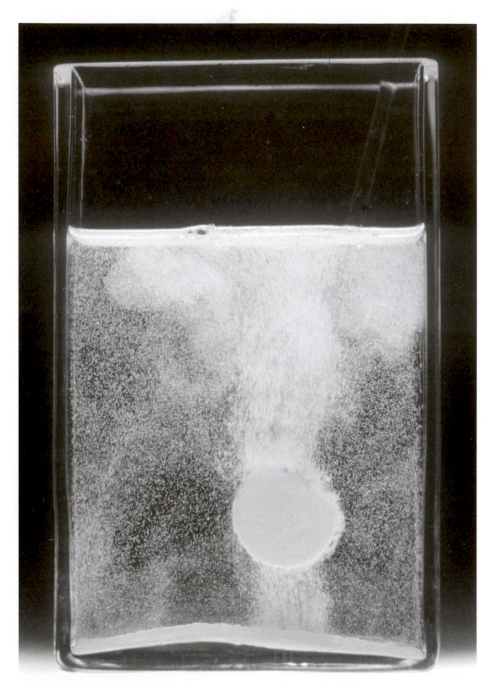

Figure 9.1.1 ► Bubbling of carbon dioxide gas as an antacid dissolves in water. From Masterton and Hurley, *Chemistry: Principles and Reactions*, 4th edition. Orlando: Harcourt, 2001. Photo courtesy of Charles D. Winters.

Figure 9.4.1 ▶ The blue and pink colors of hydrangeas that result from growing in soil with different degrees of acidity. From Masterton and Hurley, *Chemistry: Principles and Reactions*, 4th edition. Orlando: Harcourt, 2001. Photo courtesy of Charles D. Winters.

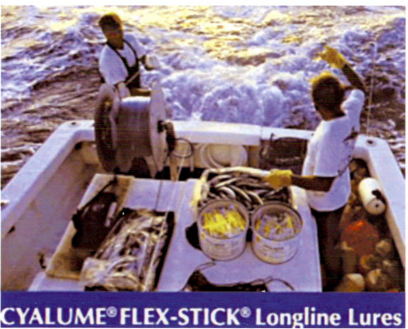

CYALUME® FLEX-STICK® Longline Lures

Figure 12.1.1 ▶ The use of chemiluminescent lures in the commercial long line fishing industry to attract such deep sea fish as tuna and swordfish. Photo from Omniglow Corporation.

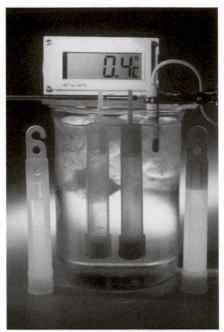

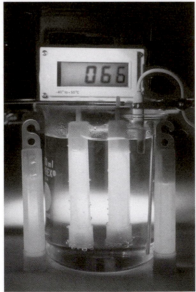

Figure 12.1.2a & b ▶   Cyalume® lightsticks immersed in water at a low (0.4°C) and a higher (6.6°C) temperature to illustrate the temperature dependence of the rate of the chemical reaction that generates the luminescent glow. The intensity of the light sticks increases as the temperature is raised as the chemical reaction rate is larger at the higher temperature. From Masterton and Hurley, *Chemistry: Principles and Reactions*, 4th edition. Orlando: Harcourt, 2001. Photo courtesy of Charles D. Winters.

Figure 12.2.1 ▶ Astronaut Eugene A. Cernan, Commander of the Apollo 17 mission, is photographed by Astronaut Harrison H. Schmitt, whose photo is reflected in the gold visor. (Courtesy NASA Scientific and Technical Program.)

Figure 12.3.1 ▶ A rare high-altitude red aurora created in the ionosphere. Photo by Jan Curtis, courtesy of The Exploratorium.

Figure 12.3.2 ▶ The aurora borealis over northwestern Canada. From Masterton and Hurley, *Chemistry: Principles and Reactions*, 4th edition. Orlando: Harcourt, 2001. Photo courtesy of George Lepp/Tony Stone Images.

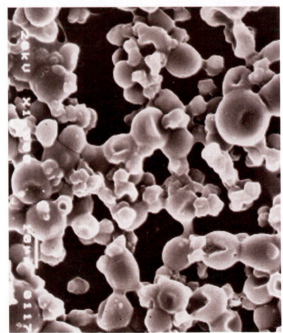

Figure 14.2.1 ▶ A scanning electron micrograph, magnified 1000 times, illustrating the morphology of the particles in a pressure-sensitive scratch-and-sniff ad. Courtesy of Museum of Science, Boston.

some drain cleaners rely on the heat and vigorous bubbling produced from the action of strong base on finely divided particles of aluminum metal.[1,2] The overall reaction liberates hydrogen gas as follows:

Dissolution of strong base in water:

$$NaOH\ (s) \rightarrow Na^+\ (aq) + OH^-\ (aq).$$

Reaction of hydroxide ion with aluminum to form a soluble aluminum-containing ion and hydrogen gas:

$$2\ Al\ (s) + 6\ H_2O\ (l) + 2\ OH^-\ (aq) \rightarrow 2\ Al(OH)_4^-\ (aq) + 3\ H_2\ (g).$$

The soluble ion $Al(OH)_4^-$ (aq) is known as the *aluminate* ion. The vigorous evolution of hydrogen gas helps to physically dislodge undissolved grease particles from the walls of plumbing.

KEY TERMS: **dissolution** **caustic** **alkaline**

References

[1] "Material Safety Data Sheet: Drano." Drackett Products Company, http://www.sanitarysupplyco.com/m74610050.txt

[2] "Material Safety Data Sheet: Liquid Drano." Drackett Products Company, http://www.biosci.ohio-state.edu/~jsmith/MSDS/DRANO%20LIQUID.htm

[3] "Material Safety Data Sheet: Liquid-Plumr." Clorox Company, http://www.biosci.ohio-state.edu/~jsmith/MSDS/LIQUID-PLUMR.htm

[4] "What's the Shiny Stuff in Drano." John T Wood and Roberta M. Eddy, J. Chem. Educ. 1996, 73, 463, http://jchemed.chem.wisc.edu/Journal/Issues/1996/May/abs463.html

[5] "Reference Data Sheet for Chemical and Enzymatic Drain Cleaners." William D. Sheridan, http://www.meridianeng.com/draincle.html

Chapter 10 ▶

Connections
to Thermodynamics

10.1 Why Do CAT Scans Often Cause a "Warm Flush" Sensation?

For the optimal viewing of soft tissues and organs by CAT scans, patients are often administered (either orally or intravenously) certain substances known as contrast media. While these substances have the vital function of enhancing the differences in tissue density for the production of superior X-ray images, some minor side effects are often reported. Interestingly, it is the chemical nature of these contrast media that not only dictates their capacity for amplifying X-ray images but also triggers the reaction of patients to the media.

The Chemical Basics

Patients often report a warm, flushed feeling as one of the side effects of injection of a contrast medium in preparation for a CAT scan. What causes this rise in body temperature? The answer lies in the ionic nature of the contrast medium and in the body's attempt to regulate the concentrations of all substances in the blood.

Introduction of a water-soluble ionic substance into the vascular system results in an increase in the number of particles in the bloodstream as the contrast substance dissolves. The body possesses several internal regulation systems and, when perturbed by an injection, attempts to restore the concentrations of substances in the blood to their "normal" or preinjection levels. To re-equilibrate the system, water from the cells of surrounding body tissue moves into the blood plasma through capillary membranes. This transfer of water is an example of *osmosis*, the diffusion of a solvent (water) through a semipermeable membrane (the blood vessels) into a more concentrated solution (the blood) to equalize the concentrations on both sides of the membrane. To accommodate the increase in water volume, the blood vessels must dilate or expand in size. An increase in the volume of circulating blood (due to the added water) causes an extra expenditure of energy by the body, producing the flushed or warm feeling. As this effect is a consequence of the process of osmosis, the effect is referred to as *osmotoxicity*. Side effects also associated with the increased blood volume and flow include excessive thirst and enhanced renal excretion.

The Chemical Details

Some of the ionic iodinated substances used in CAT scans include amidotrizoates (salts of diatrizoic acid) and iothalamates (salts of iothalamic acid).[1] The typical cations are sodium ion or meglumine (*N*-methyl-*D*-glucamine) ion (Fig. 10.1.1),

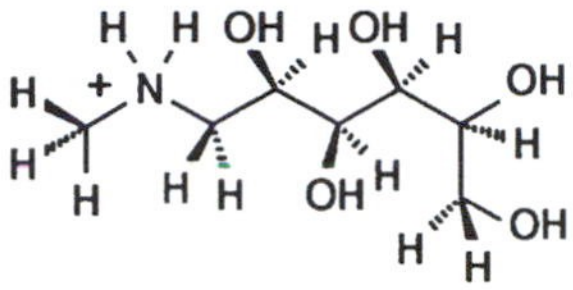

Figure 10.1.1 ▶   The molecular structure of the meglumine (*N*-methyl-*D*-glucamine) cation.

and diatrizoate (Fig. 10.1.2) and iothalamate (Fig. 10.1.3) are the common anions. These water-soluble substances have a relatively high iodine content relative to the organic portion of the substance. For example, iotalamic acid has a molecular formula of $C_{11}H_9I_3N_2O_4$ and a corresponding molecular weight of 613.92 g mol^{-1}. On a percent weight basis, iotalamic acid contains 62.0% iodine. A high iodine concentration is desirable to enhance the absorption of X-rays while only introducing a small amount of the contrast agent into the bloodstream and body tissue.

KEY TERMS: *osmosis*   *dissolution*

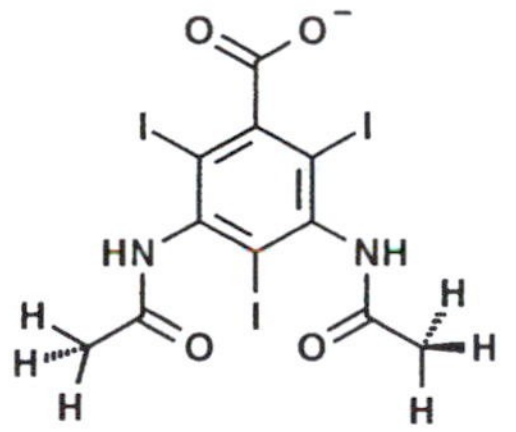

Figure 10.1.2 ▶   The molecular structure of the diatrizoate anion.

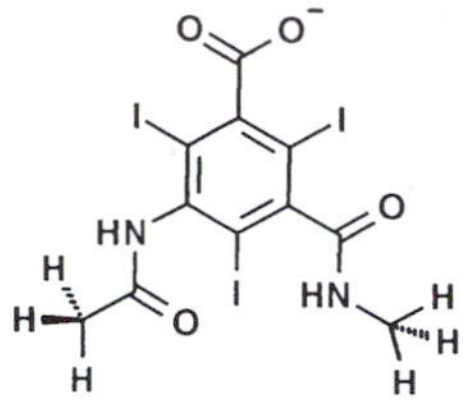

Figure 10.1.3 ▶ The molecular structure of the iothalamate anion.

References

[1] "Contrast Media," Dr. Barber, Radiology, VA-MD Regional College of Veterinary Medicine, http://education.vetmed.vt.edu/Curriculum/VM8544/barber/07/index.html

Related Web Sites

▶ "Iothalamate Sodium, Conray-400," Mallinckrodt Material Safety Data Sheet, http://imaging.mallinckrodt.com/_Attachments/MSDS/connr.htm

▶ "Hypaque sodium (sodium diatrizoate)," Material Safety Data Sheet, LaMotte Company, www.lamotte.com/WEB-SITE/SIDEFRAM/MSDS/4171.PDF

10.2 What Provides the Insulating Qualities in Double- and Triple-Pane Windows?

Thermal pane windows are recommended for better insulation to reduce energy costs. What chemistry is used in this energy-efficient cutting-edge technology?

The Chemical Basics

Particular coatings are incorporated in double- and triple-pane glass that allow the passage of visible light but selectively block infrared radiation. This two-way block prevents interior heat from escaping on cold days and impedes the heat build-up on warm days. The insulation value of such windows is also often enhanced by sandwiching a layer of an inert gas such as argon, krypton, and/or sulfur hexafluoride between the panes of glass. These gases have lower thermal conductivities than the oxygen and nitrogen molecules in air and are thus poorer transmitters of heat (see Tables 10.1 and 10.2).

The Chemical Details

What physical properties of gas molecules influence their ability to conduct heat? A quantitative dependence of the thermal conductivity of a gas can be expressed in terms of the gas molecules' specific heat, mass, and cross-sectional area. Specif-

TABLE 10.1 ▶ Gas Thermal Conductivities as a Percent of the Conductivity of Air at 1 atm and 300 K [1]

	Air	H_2	He	Ar	CO_2	SF_6	CF_4
K_{gas}/K_{air}	100%	695%	563%	67%	62%	50%	62%

TABLE 10.2 ▶ Absolute Gas Thermal Conductivities at 1 Bar [1]

	Air	H_2	He	Ar	CO_2	SF_6	CF_4
K_{gas} / cal/cm K s	6.18	43.5	36.3	4.23	3.87	3.33	4.06

ically, gas thermal conductivity is directly proportional to the specific heat C_V of the gas molecules and inversely proportional to the square root of the mass m and the cross-sectional size σ of the molecules[1]:

$$K_{gas} \propto \frac{C_V}{\sigma \sqrt{m}}.$$

KEY TERMS: heat capacity

References

[1] "Why Argon." Eric Maiken, http://www.cisatlantic.com/trimix/emaiken/Argon.htm

Related Web Sites

▶ "Energy Efficient Windows." Energy Efficiency and Renewable Energy Network, U.S. Department of Energy, http://www.eren.doe.gov/erec/factsheets/eewindows.html

10.3   How Does a Snow-Making Machine Work?

Great skiing depends on sophisticated and complex snow management by ski resorts. While fresh, dry snow is desired by ski connoisseurs, resort operators know that dense, wet snow covers terrain more thoroughly and creates a much more durable base for high traffic areas. How has science aided Mother Nature in enhancing and extending the winter season with reliable snow-making procedures?

The Chemical Basics

A snow-making machine contains a mixture of compressed air and water vapor. The machine is operated by quickly expelling air and water vapor to cause a rapid expansion in volume due to the large pressure difference between tank and atmosphere. The energy needed to expand the compressed air and water vapor is derived from these gases, lowering their temperature. The cooling effect leads to the freezing of water vapor as a solid that we view as snow.

The Chemical Details

A snow-making machine is a practical example of the First Law of Thermodynamics at work. The rapid expulsion of the mixture of compressed air and water vapor at high pressure (typically about 20 atm) ensures that the process is adiabatic; that is, no heat is transferred from or lost to the surroundings during the rapid expansion. For an adiabatic process, $q = 0$. Thus, by the First Law, the change in the internal energy of the gases is equal to the work performed by the gases as they expand:

$$\Delta E = q + w$$
$$\Delta E = w \qquad (\text{since } q = 0).$$

The sign of the work as the compressed gases expand is negative:

$$\Delta V > 0 \qquad \text{and} \qquad w = -P_{\text{ext}} \Delta V.$$

For an ideal gas, a relationship exists between the change in internal energy during a given process and the change in temperature for the gas during that process: $\Delta E = nC_V \Delta T$. Since $w < 0$, then $\Delta E < 0$, and thus ΔT must be negative. This cooling effect lowers the temperature of the air, and the cold air freezes the water vapor into snow.

KEY TERMS: **First Law of Thermodynamics**   **adiabatic process**   **heat**

work   **internal energy**

Related Web Sites

▶ Making Snow, Mary Bellis, About, Inc.,
   http://inventors.about.com/library/inventors/blsnow.htm

▶ "Snowmaking 101." Ratnik Industries, http://www.ratnik.com/snowmaking.html

Why Does a Plaster Cast Warm as the Cast Hardens?

Plaster casts are often used to hold an injured joint or bone in place to facilitate healing. As the plaster cast is applied, the patient often notes that the plaster cast feels warm. Chemistry provides an explanation for this observation.

The Chemical Basics

The plaster cast forms through a chemical reaction upon the addition of water to Plaster of Paris. This reaction releases heat, warming the hardened cast.

The Chemical Details

When Plaster of Paris, $CaSO_4 \cdot \frac{1}{2}H_2O$ (s), is mixed with water, solid gypsum, $CaSO_4 \cdot 2 H_2O$ (s), is formed in an exothermic reaction, that is, a heat-releasing reaction. Thus, as the cast hardens, the plaster cast becomes warm:

$$CaSO_4 \cdot \frac{1}{2}H_2O \text{ (s)} + \frac{3}{2}H_2O \text{ (l)} \rightarrow CaSO_4 \cdot 2 H_2O \text{ (s).}$$

We can calculate the value of ΔH°_{rxtn} for the cast-formation reaction at 298 K using the enthalpies of formation of reactants and products:

$$\Delta H^\circ_{reaction} = \sum \Delta H^\circ_f \text{ (products)} - \sum \Delta H^\circ_f \text{ (reactants)}$$

$$\Delta H^\circ_{reaction} = \Delta H^\circ_f \text{ (CaSO}_4 \cdot 2H_2O, \text{ s)} - \Delta H^\circ_f \left(CaSO_4 \cdot \frac{1}{2}H_2O, \text{ s}\right)$$

$$- \frac{3}{2} \Delta H^\circ_f \text{ (H}_2O, \text{l)}$$

$$\Delta H^\circ_{reaction} = -2021.1 \text{ kJ mol}^{-1} - (-1575.2 \text{ kJ mol}^{-1})$$

$$- \frac{3}{2} \left(-285.83 \text{ kJ mol}^{-1}\right)$$

$$\Delta H^\circ_{reaction} = -17.2 \text{ kJ mol}^{-1}.$$

As a sample calculation, we could estimate how hot the cast might possibly get if the temperature when the doctor applied the cast were initially 25°C. The cast would be the hottest if none of the heat from the cast-formation reaction were lost to the surroundings. All of the heat released by the reaction would be used to heat the products. As an approximation, the reaction could be assumed to occur completely with pressure remaining constant with both temperature-independent ΔH°_{rxtn} and heat capacity values. No actual weight of cast is needed for the calculation.

Thus,

$$\Delta H_{\text{rxtn}}^{\circ} = \Delta H^{\circ} \text{ (heating cast)}$$
$$n\ 17.2 \text{ kJ mol}^{-1} = nC_{\text{P}}\Delta T = n(186.2 \text{ J K}^{-1}\text{mol}^{-1})\Delta T$$
$$\Delta T = 92.4 \text{ K}$$
$$T_{\text{final}} = 25\,^{\circ}\text{C} + 92\,^{\circ}\text{C} = 117\,^{\circ}\text{C} \qquad \text{Ouch!!}$$

Because the skin is not burned when a cast is applied, some of these assumptions must not hold! In particular, heat must be lost to the surroundings. Furthermore, the reaction probably does not go to completion or proceeds over a time interval that allows the heat to be generated in the cast gradually. Excess water (not just the stoichiometric amount required to balance the reaction) is used when preparing the cast; the excess water serves to absorb some of the heat generated from the reaction as well. Moreover, all of the heat is not accumulated in a single point in the cast—a large surface area leads to a greater dispersal of heat. Other assumptions are probably also not strictly true (e.g., temperature-independent heat capacity and enthalpy of reaction), but these assumptions are not likely to dramatically affect the predicted final temperature. Heat loss to the surroundings is thus likely an important process that prevents the final temperature of the cast from reaching its maximum potential value.

KEY TERMS: **enthalpy of reaction**   **heat capacity**   **enthalpy of formation**
adiabatic

10.5   What Causes an Instant Ice-Pack to Cool?

Most first-aid kits contain instant cold compresses to treat injuries and burns. A simple chemical reaction, initiated by the user, provides for the immediate application of a cooling remedy.

The Chemical Basics

To provide cold therapy for cuts, bruises, sprains, and lacerations, an instant ice pack for first-aid treatment uses a chemical reaction that requires heat in order to occur. An instant Ice-Pack contains two compartments—one containing liquid water, the other a solid. The pack is activated by squeezing the liquid compartment to break an inner seal that permits the mixing of the two compartments. Heat is withdrawn from the surroundings by the reacting chemicals, lowering the temperature of the ice-pack contents.

The Chemical Details

An instant ice pack for first-aid treatment uses the endothermic nature of the dissolution of an ionic salt in water to provide cold therapy. Two typical materials that absorb heat as they dissolve in water are ammonium nitrate and ammonium chloride:

$$NH_4NO_3 \text{ (s)} \rightarrow NH_4^+ \text{ (aq)} + NO_3^- \text{ (aq)}$$
$$NH_4Cl \text{ (s)} \rightarrow NH_4^+ \text{ (aq)} + Cl^- \text{ (aq)}.$$

The standard enthalpy of reaction for the dissolution of ammonium nitrate in water can be calculated using the enthalpies of formation for reactants and products:

$$\Delta H_{\text{reaction}}^{\circ} = \sum \Delta H_f^{\circ} \text{ (products)} - \sum \Delta H_f^{\circ} \text{ (reactants)}$$
$$\Delta H_{\text{reaction}}^{\circ} = \Delta H_f^{\circ} \text{ (NH}_4^+, \text{ aq)} + \Delta H_f^{\circ}(\text{NO}_3^-, \text{ aq)} - \Delta H_f^{\circ} \text{ (NH}_4\text{NO}_3, \text{s)}$$
$$\Delta H_{\text{reaction}}^{\circ} = -132.51 \text{ kJ mol} + (-205.0 \text{ kJ mol}) - (-365.56 \text{ kJ mol})$$
$$\Delta H_{\text{reaction}}^{\circ} = +28.1 \text{ kJ mol}.$$

In a similar fashion, the enthalpy of reaction for the dissolution of ammonium chloride in water can also be determined:

$$\Delta H_{\text{reaction}}^{\circ} = \sum \Delta H_f^{\circ} \text{ (products)} - \sum \Delta H_f^{\circ} \text{ (reactants)}$$
$$\Delta H_{\text{reaction}}^{\circ} = \Delta H_f^{\circ} \text{ (NH}_4^+, \text{ aq)} + \Delta H_f^{\circ}(\text{Cl}^-, \text{ aq)} - \Delta H_f^{\circ} \text{ (NH}_4\text{Cl}, \text{s)}$$
$$\Delta H_{\text{reaction}}^{\circ} = -132.51 \text{ kJ mol} + (-167.2 \text{ kJ mol}) - (-314.43 \text{ kJ mol})$$
$$\Delta H_{\text{reaction}}^{\circ} = +14.8 \text{ kJ mol}.$$

Thus, the enthalpy change is about half that for the dissolution of ammonium nitrate in water. An estimate of the temperature achieved by the ice pack can be calculated knowing the amounts of water and salt present and assuming that the salt dissolves completely.

For example., suppose a sample instant ice pack contains 75.0 mL of water and 25.0 g of NH_4NO_3. We will assume that all 25.0 g of NH_4NO_3 dissolves in the liquid water (Remember that NH_4^+ salts are soluble, although the solubility may be limited with the small volume of water present!):

$$\# \text{ mol of } NH_4NO_3 = 25.0 \text{ g}/(80.04 \text{ g/mol}) = 0.312 \text{ mol}.$$

Then

$$\Delta H_{\text{reaction}}^{\circ} = 0.312 \text{ mol} \times 28.1 \text{ kJ mol} = 8.76 \text{ kJ}.$$

We can make an assumption that all of the heat absorbed at constant pressure ($\Delta H_{\text{reaction}}^{\circ} = q_P$) comes from the water and leads to cooling of the 75.0 mL of water. We must remember to change the sign of $\Delta H_{\text{reaction}}^{\circ}$ or q_P, as the heat is transferred from the water to the ammonium nitrate. Finally, we note that our calculation assumes that no heat is transferred to the water from the surroundings. Assuming a density of water equal to 1.00 g/mL, there are 75.0 g of water:

$$n(H_2O) = 75.0 \text{ g}/(18.0 \text{ g/mol}) = 4.17 \text{ mol}$$
$$C_P(H_2O, l) \text{ at } 25°C = 75.29 \text{ J/K} \cdot \text{mol}$$
$$\Delta H°_{\text{reaction}} = nC_P\Delta T = (4.17 \text{ mol})(75.29 \text{ J/K mol})\Delta T$$
$$= -8.76 \text{ kJ}$$
$$\Delta T = -27.9 \text{ K}$$

Assuming that we start at 298 K, the final T will be around 270 K or $-3°C$ or $25°F$. Brrr!!! Our calculation has also involved a number of other assumptions, including that we have assumed a temperature-independent enthalpy of reaction and a temperature-independent heat capacity for the water. We have also assumed that the water does not freeze (would release some heat). Nevertheless, the calculation gives a fairly reasonable estimate of the temperature drop that provides the cooling therapy of an instant ice pack.

KEY TERMS:	endothermic reaction	enthalpy of reaction
	dissolution	enthalpy of formation

Other Questions to Consider

2.4   Why do cosmetic cold creams feel cool when applied to the skin? *See* p. 9.

5.3   Why is fighting magnesium fires with water or CO_2 dangerous? *See* p. 39.

5.9   Why are bombardier beetles known as "fire-breathing dragons"? *See* p. 49.

5.17   Why does baking soda extinguish a fire? *See* p. 63.

5.19   What causes puff pastry to expand? *See* p. 67.

8.2   Why do citrus growers spray their trees with water to protect them from a freeze? *See* p. 107.

9.6   How do Drano and Liquid-Plumr unclog drains? *See* p. 126.

Chapter 11 ▶

Connections to Kinetics

11.1 Why Do Some Batteries Last Longer When Stored in a Refrigerator?

Does it really help to store batteries in a refrigerator?

The Chemical Basics

The conversion of chemical energy of a cell/battery into electrical energy is known as the process of discharge. All batteries have characteristics that lead to a slow loss of charge over time (known as battery self-discharge). Additionally, some battery components can deteriorate over time (such as occurs with the corrosion of zinc in zinc–carbon batteries) or volatilize (such as occurs with the loss of water through evaporation). As any of the reactions are characterized by a reaction rate, then the rate constant for the reaction can exhibit a temperature dependence. If the reaction mechanism involves a single rate-determining step, then the rate constant for that elementary step will behave according to the Arrhenius law. This law gives a quantitative expression to relate the increase in a reaction rate with higher temperature. In general, the higher the storage temperature, the worse the capacity retention. Thus, for some batteries, self-discharge and deterioration reactions can be slowed considerably by storing the battery at low temperature when not in use. Nickel metal hydride (NiMH) batteries have a self-discharge rate of only 0.8%/day at 20°C but a self-discharge rate of 6%/day at 45°C.[1] The shelf-life of primary (nonrechargeable) batteries such as the zinc–carbon or alkaline-manganese systems[2,3] can be significantly extended by storage at lower temperatures.

The Chemical Details

In 1887 the Swedish chemist Svante Arrhenius suggested that before a chemical reaction could occur, there must be some minimum kinetic energy possessed jointly by the molecules undergoing collision. If (two) molecules collide with less than this critical amount of energy, they will recoil without undergoing chemical change. If this added potential energy were greater than some critical amount known as the "activation energy (E_a)," then reactants proceed to products. Initially, Arrhenius empirically derived the temperature dependence of rate constants.

His relationship of the rate constant k with temperature T in Kelvin involved a constant A known as the pre-exponential factor and the activation energy E_a:

$$k = Ae^{-E_a/RT} \qquad \ln k = \ln A - E_a/RT.$$

KEY TERMS:	Arrhenius equation	reaction rate
	activation energy	rate constant

References

[1] "Inside the Battery: An Electrochemical Background." IMREs Lapcenter, http://ural2.hszk.bme.hu/~ki023/psibatt/insidethebattery.htm

[2] Rayovac Battery Solutions, http://www.rayovac.com/busoem/oem/specs/clock5b.shtml

[3] "Self-Discharge of Batteries as a Function of Temperature." http://www.corrosion-doctors.org/Batteries/self-compare.htm

Other Questions to Consider

12.1   Why do lightsticks glow? *See* p. 139.

14.1   How does a timed-released medicine work? *See* p. 208.

Chapter 12

Connections to Light

12.1 Why Do Lightsticks Glow?

Few of us recognize that the fascinating glow of a lightstick at a carnival or fair arises from a complex series of chemical reactions. The chemist's palette — a variety of ingredients with specific chemical structures — produces the rainbow of colors seen in these novelties.

The Chemical Basics

Lightsticks and other glow-in-the-dark products involve a phenomenon similar to the processes that lead to a firefly's glow or a lightning strike. This phenomenon is called *chemiluminescence* — the process whereby chemical energy produced by a chemical reaction is transformed into light energy, a different form of energy. In a firefly the light-producing chemical reaction is triggered by a catalyst known as an enzyme. In a lightning storm, an electrical discharge in the atmosphere sets off a sequence of reactions that leads to the flash of light in the sky. In a lightstick, when you follow a manufacturer's instructions to bend, snap, and shake the stick, you initiate a chemical reaction by mixing the substances contained within separate compartments in the plastic tube. Energy generated during the course of the reaction is accepted by a dye molecule contained within the lightstick and then released in the form of colored light with no accompanying heat.

The Cyalume lightsticks marketed by Omniglow Corporation have found an extensive range of applications beyond the initial toy and novelty market. The intense yet cool light provided by a chemical lightstick is ideal for emergency lighting, traffic control, and hazard identification. Numerous sports products, including golf balls, footballs, hockey pucks, badminton birdies, and wiffle balls, use replaceable lightsticks to create innovative nighttime activities. In virtually every amusement park open past sunset, we can find varieties of glow-in-the-dark colored necklaces, bracelets, and bands. Military covert, nighttime, and emergency operations have been enhanced with the use of visible and infrared chemiluminescent products. The United States and Allied forces during Operation Desert

Storm employed lightsticks for underwater operations, nighttime personnel and ship-to-ship identification, hazard demarcation, and color-coding of units, vehicles, and equipment. Commercial fishing fleets have found deep sea fish such as tuna and swordfish to be attracted to the light emitted by lightsticks when the fish swim in surface waters at night. In particular, the longline fishing industry utilizes lightsticks to illuminate their monofilament long lines, often extending in length up to 80 miles with more than 3000 hooks (see color Fig. 12.1.1). With the variations in color, intensity, and duration of light emission possible, numerous applications in other commercial industries are likely.

The Chemical Details

The color, intensity, and duration of light that we observe depend upon the exact composition of a lightstick. Let's explore how the chemical composition affects these properties of the lightstick. The amount of energy produced by the chemical reaction between the original components in the lightstick dictates the color of the glow. Certain chemical reactions generate large amounts of energy and give rise to short-wavelength light (i.e., energy $\propto$ 1/wavelength) that is blue or violet in color. Other reactions are somewhat less energetic and give rise to red and orange glows at the longer wavelength end of the visible spectrum. The efficiency of the conversion of energy from the chemical reaction into light is one factor that affects the intensity of the light emission. The more efficient the conversion, the brighter the lightstick's glow. In addition to a characteristic energy output, each chemical reaction has its own efficiency.

The intensity of a lightstick is also sensitive to temperature (see color Fig. 12.1.2). An increase in temperature accelerates the rate of the chemical reaction, increasing the intensity or amount of light that can be generated from the dye molecule in a given time period. By the same token, you can prolong the life of a lightstick by storing it in the freezer. At lower temperatures, the chemical reaction slows down, producing less energy per unit time and yielding less intense light that persists for a longer period of time. By making light intensity measurements at a variety of temperatures, adherence to the Arrhenius law can be demonstrated (i.e., intensity $\propto$ rate $\propto e^{-1/T}$). The duration of the lightstick's glow is dictated by the amount of chemical reactants contained in the tubing. At any temperature, once all of the original chemicals contained within the lightstick are consumed by the reaction, chemiluminescence is no longer possible, and the lightstick fades to darkness.

Let's look at a general chemical reaction for the chemiluminescence of a lightstick in some detail. Hydrogen peroxide, H_2O_2, is the primary ingredient contained in an aqueous solution within an inner thin-glass ampule within the lightstick. By bending the plastic tubing of the lightstick, the thin vial of H_2O_2 is broken, mixing H_2O_2 with a surrounding solution of a phenyl oxalate ester $((COOC_6H_5)_2)$ and a fluorescent dye. The ester and peroxide react in a series of several steps to generate a highly energetic C_2O_4 intermediate as in Fig. 12.1.3.[1]

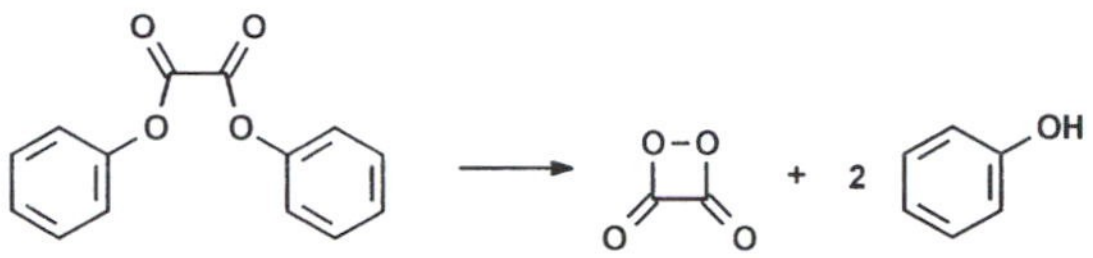

Figure 12.1.3 ▶   The reaction of phenyl oxalate ester and hydrogen peroxide to generate a highly energetic C_2O_4 intermediate.

Variations in the substitution on the phenyl rings of the oxalate ester primarily affect the yield of the C_2O_4 intermediate. The energetic nature of the intermediate presumably arises from the significant strain imposed by the four-membered ring. Recall that the ideal bond angle around a planar sp^2-hybridized carbon is 120°, not 90°. The *chemielectronic* step — the step in which the *chemical* energy of the intermediate is converted into *electronic* energy in the fluorescent dye — may be written as in Fig. 12.1.4.

Figure 12.1.4 ▶   A chemielectronic step, i.e., a step in which the chemical energy of an intermediate is converted into electronic energy in a fluorescent dye. Here the C_2O_4 intermediate releases energy as it dissociates into two carbon dioxide molecules. The energy is transferred to the accompanying fluorescent dye to generate an excited state of the dye.

The release of energy to the dye molecule or fluorescer is driven by the conformational instability of the C_2O_4 intermediate (the flat highly strained C_2O_4 prefers to be two linear CO_2 molecules). The sensitized fluorescer, denoted *fluorescer**, returns to the ground state via the emission of light:

$$\text{fluorescer}^* \rightarrow \text{fluorescer} + h\nu.$$

Emission spanning the visible and near-infrared wavelengths has been obtained through the choice of fluorescent dye. For example, 9,10-diphenylanthracene (Fig. 12.1.5) generates blue light; 9,10-bis(phenylethynyl)anthracene (Fig. 12.1.6) yields yellow-green emission with maximum output at 486 nm; rubrene (5,6,11,12-tetraphenylnaphthacene; Fig. 12.1.7) emits orange-yellow light at 550 nm; violanthrone (Fig. 12.1.8) emits orange light at 630 nm; 16,17-(1,2-ethylenedioxy)violanthrone (Fig. 12.1.9, R...R = –OCH$_2$CH$_2$O–) gives rise to a red glow at 680 nm; and 16,17-dihexyloxyviolanthrone (Fig. 12.1.9, R = –OC$_6$H$_{13}$) provides infrared luminescence at 725 nm.

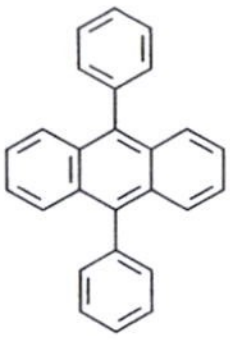

Figure 12.1.5 ► The molecular structure of 9,10-diphenylanthracene.

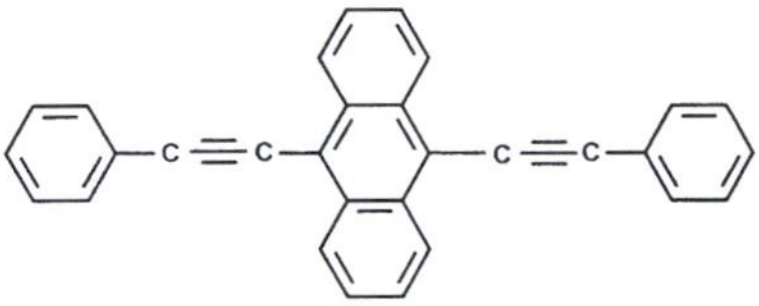

Figure 12.1.6 ► The molecular structure of 9,10-bis(phenylethynyl)anthracene.

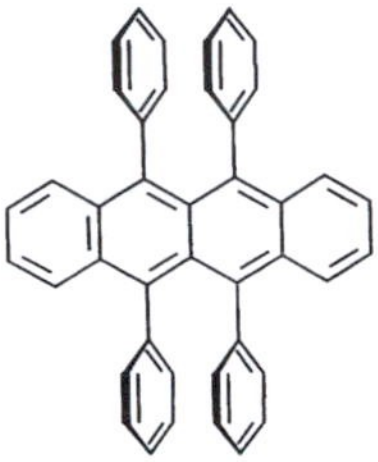

Figure 12.1.7 ► The molecular structure of rubrene (5,6,11,12-tetraphenyl-naphthacene).

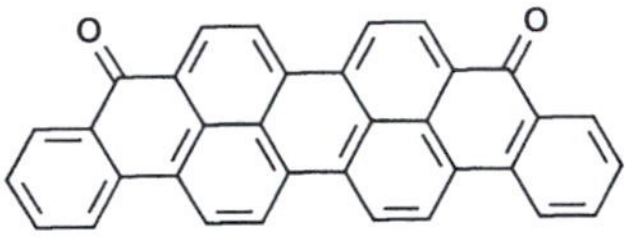

Figure 12.1.8 ► The molecular structure of violanthrone.

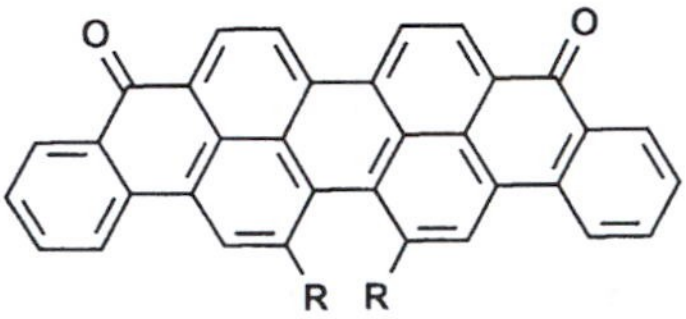

Figure 12.1.9 ► The molecular structure of 16,17-(1,2-ethylenedioxy)violanthrone with R...R = $-OCH_2CH_2O-$ and 16,17-dihexyloxyviolanthrone with each R = $-OC_6H_{13}$.

References

[1] B. Z. Shakhashiri, *Lightsticks in Chemical Demonstrations: A Handbook for Teachers of Chemistry*, Volume 1 (Madison: Univ. of Wisconsin Press, 1983), 146–152.

Related Web Sites

► "The Chemiluminescence Home Page." Dr. Thomas G. Chasteen, Sam Houston State University, http://www.shsu.edu/~chm_tgc/chemilumdir/chemiluminescence2.html

► Omniglow Corporation, http://www.omniglow.com/

► "Lightstick Chemistry." Moravian College, Chemistry Department, http://www.cs.moravian.edu/chemistry/lightstick/l_st_scheme.html

► "Lightstick Spectra." Moravian College, Chemistry Department, http://www.cs.moravian.edu/chemistry/lightstick/l_st_spec.html

12.2 Why Is an Astronaut's Visor So Reflective?

The bright reflection of the sun's rays on an astronaut's visor is keenly apparent in many of the photographs taken during spacewalks. The chemistry of the visor reveals the origin of its highly reflective nature.

The Chemical Basics

We are familiar with mirrors—polished surfaces that divert light according to the law of reflection. In ancient times, polished castings of solid tin, bronze, copper, gold, and silver served as the first mirrors. Modern mirrors consist of a plate of glass with a thin layer of aluminum or silver deposited on the backing. Thus, glass serves as the substrate—the underlying material to which a coating is applied. The brilliant white luster of aluminum and silver metal contributes to their selection as mirror coatings. While the decorative beauty of silver has been known for centuries, the *silvering* process of making mirrors was discovered in 1835 by the German chemist Justus von Liebig.

A typical mirror *reflects* both visible and near-infrared light. The thickness of the silver layer further delineates a mirror's function. Without the *transmission* of visible radiation, one cannot see through a silvered mirror. The coating on an astronaut's visor permits the visor to act as both a mirror and a transmitting shield. Coated with an ultrathin film of gold, infrared light reflects off the visor's surface while still permitting the astronaut to see visible light through the visor shield (see color Fig. 12.2.1). The relative amount of reflection to transmittance can be

controlled by the thickness of the film. A "gold-plated" visor is absolutely necessary to protect the astronaut from the infrared rays of the sun that are essentially
unfiltered in space. Even the general public can benefit from this technology —
gold-mirrored lenses are popular on some brands of sunglasses. Surfaces other
than glass have been treated with gold for infrared protection. For example, gold
tape served as a covering on the tetherlines connecting the *Gemini-Titan* 4 astro-
nauts with their spacecraft during spacewalks ("extravehicular activity"). Thin
films of gold also coat satellites in order to control the temperatures that could
result from infrared heating in space. The exterior of the canopies of F16 aircraft
are also treated with ultrathin layers of gold metal, presumably to reduce the radar
signature of the aircraft (although the purpose of the gold treatment on the cockpit
is officially classified).

The Chemical Details

An astronaut's visor is coated with thin films of the element gold. This coating
reflects up to 98% of the infrared radiation incident on the visor, protecting the
astronaut from the intensity of the sun's heat. The same principle dictates why
gold films (as thin as even 20 pm!)[1] are placed on the inside face of windows
in office buildings, reducing heat losses in winter and reflecting infrared radiation
(and thus heat) in summer. Gold is more malleable and ductile than any other
metal, enabling the creation of thin films, and its high thermal conductivity (only
silver and copper have higher thermal conductivities)[2] enables efficient cooling.
However, the high degree of reflectivity of gold in the infrared region (98%)[3]
minimizes the absorption of radiation by the gold coating deposited on the visor
or glass.

KEY TERMS:	gold	infrared radiation	reflectivity	silvering
	reflection	thermal conductivity	transmission	

References

[1] N. N. Greenwood, and A. Earnshaw, "Copper, Silver, and Gold," in *Chemistry of the Elements*, 2nd ed. (New York: Pergamon Press, 1997), Chap. 28, pp. 1173–1200.

[2] "Thermal and Physical Properties of Pure Metals." *Handbook of Chemistry and Physics*,
74th ed., ed. David R. Lide (Boca Raton, FL: CRC Press, 1993), **12**-134 to **12**-137.

[3] "Metallic Reflector Coatings, Oriel Instruments, http://www.oriel.com/netcat/VolumeIII/
pdfs/v36metct.pdf

Related Web Sites

▶ NASA Photos of Astronaut Edward H. White II during Extravehicular Activity Performed
during the Gemini-Titan 4 Space Flight,
http://science.ksc.nasa.gov/mirrors/images/images/pao/GT4/10074015.htm

▶ http://www.ksc.nasa.gov/mirrors/images/images/pao/GT4/10074016.htm

▶ http://www.ksc.nasa.gov/mirrors/images/images/pao/GT4/10074017.htm

▶ http://www.ksc.nasa.gov/mirrors/images/images/pao/GT4/10074018.htm

▶ "Gold Occurrences." R. James Weick, Geological Survey of Newfoundland and Labrador, http://www.geosurv.gov.nf.ca/education/occgold.html

12.3 Why Is the Aurora Borealis So Colorful?

And the Skies of night were alive with light, with a throbbing, thrilling flame;
Amber and rose and violet, opal and gold it came.
It swept the sky like a giant scythe, it quivered back to a wedge;
Argently bright, it cleft the night with a wavy golden edge.
Pennants of silver waved and streamed, lazy banners unfurled;
Sudden splendors of sabres gleamed, lightning javelins were hurled.
There in our awe we crouched and saw with our wild, uplifted eyes.
Charge and retire the hosts of fire in the battlefield of the skies.

—Robert W. Service, *The Ballad of the Northern Lights*

How does chemistry explain the ethereal and breathtaking Aurora Borealis?

The Chemical Basics

Have you ever been treated to the breathtaking display of the northern lights or aurora borealis of the Northern Hemisphere (or the corresponding southern lights aurora australis of the Southern Hemisphere)? The colorful auroral rings and patches of light in the nighttime sky are a wonderous sight. Aurora was the Roman goddess of dawn, and aurora borealis and aurora australis literally mean "dawn of the north" and "dawn of the south." In fact, auroras are the final event of a chain of reactions that begin with the sun. Solar flares eject energetic particles that travel throughout interplanetary space at a high velocity (400 km/s or 250 miles/s).[1] These particles are charged and can be trapped by the earth's magnetic field. As the particles travel to the earth's magnetic poles, they collide with oxygen (i.e., dioxygen, O_2) and nitrogen (i.e., dinitrogen, N_2) gases in our upper atmosphere 40–600 miles above Earth. These collisions result in a transfer of energy to the gaseous molecules that may even be sufficient to *ionize* the molecules (create a charged species by removing an electron, e.g., $O_2 + e^- \rightarrow O_2^+ + 2\,e^-$) or *dissociate* (separate) the molecules into individual atoms (e.g., $O_2 + e^- \rightarrow 2\,O^+ + e^-$). The collection of energized molecules, ions, and atoms returns to a less energetic state by releasing the acquired energy in the form of light. The color of the light observed depends on the identity of the species (see color Figs. 12.3.1 and 12.3.2).

The Chemical Details

An aurora is generated by streams of energetic charged particles (mostly electrons) that emanate from explosive activity, such as solar flares, on the surface of the sun. Interaction of the solar wind with the earth's magnetic field restricts these particles to higher latitudes. The characteristic wavelengths observed in auroras depend on the chemical identity of the energetic species created and the energy transferred from the solar particles. Highly energetic electrons can penetrate to great depths in the atmosphere, whereas low-energetic electrons only reach to the top of the thermosphere. Violet and blue light with wavelengths of 391.4 and 470.0 nm is emitted by N_2^+, predominantly located in the lower atmosphere. The sensitivity of the human eye generally does not readily detect the blue wavelengths. Crimson-red light at 630 nm is discharged by O_2^+. Both a greenish-yellow light at 557.7 nm (high energy) and a deep red light at 630.0 and 636.4 nm (low energy) are characteristic of O atoms, created via dissociation of diatomic oxygen by ultraviolet light. Yellow-green auroras arise from oxygen atoms at lower altitudes where energetic electrons penetrate. The combination of a large influx of low-energy electrons and high-altitude oxygen (200 miles up) is responsible for the rare all-red auroras.[2]

KEY TERMS:	wavelength of light	energy	emission
	ionization	dissociation	

References

[1] "The Sun." NASA Goddard Space Flight Center,
http://www-spof.gsfc.nasa.gov/Education/Isun.html

[2] "The Rare Red Aurora." Article 918, Carla Helfferich , *Alaska Science Forum*, March 22, 1989, http://www.gi.alaska.edu/ScienceForum/ASF9/918.html

Related Web Sites

▶ "Auroras: Paintings in the Sky." Mish Denlinger, The Exploratorium,
http://www.exploratorium.edu/learning_studio/auroras

▶ "The Aurora Page." Department of Geological Engineering and Sciences, Michigan Technological University, Houghton, MI, http://www.geo.mtu.edu/weather/aurora/

▶ "The Exploration of the Earth's Magnetosphere." NASA Goddard Space Flight Center,
http://www-spof.gsfc.nasa.gov/Education/Intro.html

▶ "Space Weather: A Research Perspective — The Elements of Near-Earth Space." Space Studies Board, Commission on Physical Sciences, Mathematics, and Applications, National Research Council, http://www.nas.edu/ssb/elements.html

▶ Photo of a Red Aurora, Alyeska Pipeline, http://www.alyeska-pipe.com/Photolibrary/AuroraBorealis.html

12.4   What Causes the Pearlescent Appearance of Some Paints?

Pearlescent lusters are commonly seen in many paints, inks, and cosmetics. The pearlescent pigment technology that brings us these unusual effects relies on a common mineral to achieve these opalescent qualities.

The Chemical Basics

Pearlescent pigments contain small flakes or platelets of the mineral mica that are additionally coated with a very thin layer of titanium dioxide. The simultaneous reflection of light from many layers of small platelets creates an impression of luster and sheen. By varying the thickness of the coating on the surface of the mica particles, pigment manufacturers can achieve a range of colors for the pearlescent effect.

The Chemical Details

Pearlescent pigments are derived from microscopic mica platelets ranging in size from 1 to 2 μm in thickness and up to 180 μm in diameter. Mica consists of an aluminum silicate with a crystalline structure that easily permits cleavage into very thin platelets. In fact, mica is a generic term for any one of several complex hydrous aluminosilicate minerals characterized by their platy nature and pronounced basal cleavage. The general formula for mica is $AB_{[2-3]}(Al, Si)Si_3O_{10}(F, OH)_2$.[1] Generally A represents the metal potassium (K) and B corresponds to aluminum (Al). However, some micas contain calcium (Ca), sodium (Na), or barium (Ba) for A and lithium (Li), iron (Fe), or magnesium (Mg) for B. This subclass of silicates contains rings of SiO_4 tetrahedrons linked by shared oxygens to other rings in a two-dimensional plane. This bonding arrangement produces a sheet-like structure that gives rise to flat, platy crystals with good basal cleavage.[1]

The presence of mica in pearlescent pigments only partly accounts for the appearance of the pigment. A very thin layer of the inorganic oxide titanium dioxide (TiO_2) or iron oxide (Fe_2O_3) or both is coated on the mica platelets. The various colors and pearlescent effects are created as light is both refracted and reflected from the titanium dioxide layers. The very thin platelets are highly reflective and transparent. With their plate-like shape, the platelets are easily oriented into parallel layers as the paint medium is applied. Some of the incident light is reflected

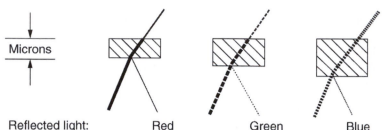

Reflected light: Red   Green   Blue

Figure 12.4.1 ▶   The multiple reflection of light from microscopic oxide layers of different dimensions leads to constructive and destructive interference of light waves, producing a particular color effect. Different thicknesses reflect different colors.

from the uppermost layers, while a portion is transmitted to lower layers and then reflected. A pearlescent luster is produced from this multiple reflection of light from many microscopic layers. Smaller mica particle size leads to smoother sheens, while larger particle size produces a higher luster or sparkled effect. In addition, the thickness of the oxide layer dictates the color observed. Multiple reflections and refractions of light lead to both constructive and destructive interference of light waves. The oxide layer thickness determines the narrow range of wavelengths of light that interfere constructively, thus producing a particular color effect. All other wavelengths of light interfere destructively and are not observed. Fig. 12.4.1 summarizes these concepts.

KEY TERMS: interference   inorganic   oxide

References

[1]   The Silicate Class, The Mineral Gallery,
      http://www.galleries.com/minerals/silicate/class.htm

Related Web Sites

▶ "Interference and Iridescent Acrylics." Golden Artist Colors, New Berlin, NY,
   http://www.goldenpaints.com/iridint.htm

▶ Microscopic photo of pearlescent paint,
   http://www.standox.de/standote/englisch/color/3schicht.htm

▶ "Understanding Vehicle Paint Technology: What Are Metallic Finishes?"
   http://www.standox.de/lack/englisch/guide3.htm#Metallic

▶ "Standothek: Design and Paint." http://www.standox.com/standote/englisch/design/
   effekt.htm

12.5 | Why Is It Incorrect to Call U.S. Bills "Paper Currency"?

Most Americans recognize the portraits that appear on the front face of U.S. currency: George Washington ($1), Thomas Jefferson ($2), Abraham Lincoln ($5), Alexander Hamilton ($10), Andrew Jackson ($20), Ulysses Grant ($50), Benjamin Franklin ($100). As with U.S. coinage, the Secretary of the Treasury, in consultation with the Commission on Fine Arts, selects the designs shown on U.S. currency. However, it is the Bureau of Engraving and Printing that is responsible for designing and printing U.S. currency at its facilities in Washington, DC, and Ft. Worth, TX. While the printed features of U.S. currency have undergone significant changes over the years, the chemistry of the "paper" has a much more stable history.

The Chemical Basics

The constant circulation of a national currency demands a durable, yet high-quality material. Beginning with the first series of U.S. bank notes issued in 1861, U.S. currency has never been printed on paper but on a cotton/linen fabric with the linen content held at $25 \pm 5\%$. Often this fabric is referred to as cotton and linen rag paper. Red and blue silk fibers from scraps and cuttings of clothing manufacturers are embedded in the cotton/linen sheet to deter counterfeiters. A commercial company, Crane & Company Inc. of Dalton, MA, has been manufacturing the rag paper since 1879, and it is illegal for anyone to manufacture, possess, or use this material or a fabric of a similar type. The history of Crane & Company is a fascinating story.[1] The association of the Crane family with U.S. currency began with the sale of paper in 1775 by Steven Crane to Paul Revere for the Massachusetts Bay Colony's first currency. Having been taught the art of paper-making by his father, Zenas Crane carried on the family business and built his first paper mill along the Housatonic River in Dalton, MA in 1801. The following advertisement appeared in the *Pittsfield Sun* in 1800[2]:

Americans—

Encourage your own manufactories, and they will Improve. Ladies save your Rags. As the Subscribers have it in contemplation to erect a Paper-Mill in Dalton, the ensuing Spring; and the business being very beneficial to the community at large, they flatter themselves they shall meet with due encouragement.

—[*Signed*] Henry Marshall,
Zenas Crane, and John Willard

As the above notice indicates, cloth rags were the basic raw material for conversion into pulp for high-quality, rag-based paper products. Although Zenas Crane retired in 1842, his sons James Brewer Crane and Zenas Marshall Crane continued the business and begin to make paper for banknotes, bonds, and securities. Two generations later in 1879, W. Murray Crane won a competition to manufacture the paper for U.S. currency. Crane & Company has held an exclusive contract with the U.S. Treasury Department to produce the specially threaded cotton and linen paper for U.S. currency. The company also manufactures paper for the foreign currencies of Canada, Mexico, Indonesia, and the Ukraine[3] in addition to its renown high-quality stationery business.

The Chemical Details

In ancient Egypt the fibers and glue-like sap of the reedy papyrus plant were the main constituents of the sheets formed for writing materials. Other woody fibers, such as mulberry, were introduced by the Chinese in the first century AD. Paper mills existed in Europe by the fourteenth century, and linen and cotton rags served as the basic raw materials through the eighteenth century. Paper mills often solicited publicly for rags because the shortage of raw materials could not keep up with the demand for paper. The manufacturing of paper from wood pulp began in 1800 to relieve the paper industry from its demand for cotton and linen rags. Even today grades of paper requiring strength, durability, permanence, and fine texture employ cotton and linen fibers derived from textile and garment mill cuttings.

One of the final steps in the paper-making process is the coating of the paper surface to achieve a variety of effects. For example, coating can enhance the uniformity of the surface for printing inks or enhance the opacity or whiteness of the paper. Titanium dioxide (TiO_2) is used to whiten and opacify all U.S. paper currency and many other forms of paper.[4] Why is titanium dioxide an optimal coating selected to whiten paper? For the human eye to perceive the color white, an object must scatter all wavelengths of light. Thus, the ability of a sheet of paper, or its coating, to scatter light will define its opacity and brightness. One mechanism of light scattering is by *refraction*, and the high refractive index of titanium dioxide makes it an effective scatterer of light. Refraction is defined as the bending of light as it passes from one medium, such as air, to another medium, such as a paper coating. The larger the difference in the refractive index of the two media, the greater the extent to which the light is bent or refracted. The extremely high refractive index values of both commonly used crystal forms of titanium dioxide — anatase and rutile — make them ideal commercial pigments for achieving brightness with low volumes of sample.[5]

KEY TERMS: refraction

References

[1] "The State of One Small Family Business: Crane & Co." Mary Furash, Inc. Online,
      http://www.inc.com/incmagazine/archives/27961141.html

[2] "Crane Museum, Dalton, Massachusetts 01226."
     http://www.berkshireweb.com/themap/dalton/museum.html

[3] "Opening up Money Supply Worries Dalton Paper Firm." Jeff Donn, Associated Press, The
     Standard-Times, http://www.s-t.com/daily/04-97/04-13-97/f03bu299.htm

[4] "Basics: Why is TiO$_2$ Used in Paper?" Millenium Chemicals,
     http://www.millenniumchem.com/Products+and+Services/Products+by+Type/
     Titanium+Dioxide+-+Paper/r_Paper+Basics/

[5] "DuPont Ti-Pure: Paper: Rutile and Anatase."
     http://www.dupont.com/tipure/paper/anarut.html

Related Web Sites

- "Crane Bro's Paper Manufacturers." http://www.holyokemass.com/historic/
  pichampden/p157.html

- "History of Paper Money." Federal Reserve Bank of San Francisco,
  http://www.frbsf.org/federalreserve/money/funfacts.html#A3

- "Fundamental Facts about U.S. Money." Federal Reserve Bank of Atlanta,
  http://www.atl.frb.org/publica/brochure/fundfac/money.htm

- "Money Hot Off the Press." Bureau of Engraving and Printing, United States Treasury, A
  30-second video capsule of how currency is printed, http://www.treas.gov/bep/money.mpg

- "The Making of Money." Bureau of Engraving and Printing, United States Treasury, A
  39-second video with sound on the currency-making process, http://www.treas.gov/bep/
  making.mpg

- "Anatomy of a Bill: The Currency Paper, Secrets of Making Money." NOVA Online,
  http://www.pbs.org/wgbh/nova/moolah/anatomypaper.html

- "Money Making." The AFU and Urban Legend Archive,
  http://www.urbanlegends.com/misc/money_making.html

- "Six Kinds of United States Paper Currency." Kelley L. Ross, Ph.D.,
  http://www.friesian.com/notes.htm

- "A Brief History of Our Nation's Paper Money." 1995 Annual Report, Federal Reserve
  Bank of San Francisco, http://www.frbsf.org/publications/federalreserve/annual/
  1995/history.html

- "Treasury Home Page: Existing Contracts: Massachusetts."
  http://www.treas.gov/sba/ma.html

- "DuPont Ti-Pure Titanium Dioxide." http://www.dupont.com/tipure/paper/index.html

- "The Mineral Rutile." Amethyst Galleries, Inc.,
  http://mineral.galleries.com/minerals/oxides/rutile/ rutile.htm

- "The Mineral Anatase." Amethyst Galleries, Inc.,
  http://mineral.galleries.com/minerals/oxides/anatase/anatase.htm

Other Interesting References

▶ "Zenas Crane," in *Encyclopedia of Entrepreneurs*, ed. Anthony Hallett and Diane Hallett (New York: Wiley, 1997), 131–132.

12.6 What Is the Purpose of the Thread That Runs Vertically through the Clear Field on the Face Side of U.S. Currency?

The Bureau of Engraving and Printing in Washington, DC, is responsible for the design, engraving, and production of U.S. banknotes. Several of the special design features evident in U.S. currency today take advantage of technological advancements in the ink and paper industries. The highly sophisticated chemistry of such materials is an effective deterrent against counterfeiting.

The Chemical Basics

Have you ever examined the details on paper U.S. currency? The denomination, portrait, and back design are commonly recognized. The Treasury seal, Federal Reserve seal, signature of the U.S. Treasurer, motto, and serial numbers are also clearly identified. Some design features are not as readily apparent. One such feature introduced in 1990 series currency is a clear thread embedded in paper currency as a security measure against counterfeiting. The security thread is a polyester thread on which a denomination identifier is printed. Both the thread and the printing are visible only with a light source and can be viewed from either the face or the back of the note. For the two highest denomination bills, $100 and $50, the security thread repeats the wording "USA 100 USA 100" and "USA 50 USA 50," respectively. For the next three lower denominations ($20, $10, and $5), the printing consists of the abbreviation "USA" followed by the written denomination in capital letters, as in "USA TEN," repeated along the length of the thread. No security thread is included in the $1 note. Beginning with the 1996 series notes, the placement of the thread varies, with the position indicative of the bill's denomination. In addition, the thread on the new $100 note first issued on March 25, 1996 was redesigned to glow or fluoresce a red color when exposed to ultraviolet light in a dark setting. The thread is positioned to the immediate left of the oval frame on Ben Franklin's portrait. The new $50 notes first issued on October 27, 1997 also have the added feature that the security thread, now to the right of the portrait of President Grant, glows yellow when exposed to ultraviolet light in a dark environment. The printing on the thread also includes a flag in addition to "USA 50." The redesigned $20 bill was unveiled in May 1998 and put into circulation in the fall of that year. The vertical thread is embedded to the far left of President Jackson's portrait (to the left of the Federal Reserve Seal). The words "USA TWENTY" and a flag can be seen from both sides against a light. The number "20" appears in the star field of the flag. The thread fluoresces green under an ultraviolet light. The design features for the new $10 and $5 notes, released in May 2000, include security threads that glow orange and blue, respectively, when held under an ultraviolet light. The introduction of the new security measures has not always occurred without error. In November 1996 officials at the Bureau of Printing and Engraving of the Treasury Department discovered that the positions of the polymer security thread and the watermark on $4.6 million worth of the newly designed $100 bills were swapped — the thread incorrectly appeared on the right side of Benjamin Franklin's portrait and the watermark on the left.[1] These misprinted bills are still legal tender, yet they are likely to be worth much more than their face value to collectors. You might wonder why the U.S. Treasury is varying the position of the security thread in its currency. Counterfeiters often attempt to "raise notes," that is, bleach out the paper of a low denomination and reprint a higher denomination onto the authentic paper. With the position of the security thread constrained to the bill's denomination, this type of fraud will be easily detected.

The Bank of Latvia (Latvijas Banka) also incorporates invisible fluorescent fibers in their currency as a security measure. As with the security thread on U.S. bank notes, the Latvian thread fluoresces under ultraviolet light, emitting three different colors. In addition, the printing ink employed for the red serial numbers located in the upper center portion of the face of the notes fluoresces upon exposure to ultraviolet light. Singapore currencies include specially formulated inks to fluoresce under ultraviolet light, including black serial numbers that glow green and a red seal of the Minister of Finance that emits orange illumination. Fluorescent inks are employed for denomination figures, seals, and other objects in Hong Kong dollars, Malaysian dollars, Taiwan yuan, and Indonesian rupiahs. Special fluorescent fibers were incorporated in the paper-making process for Deutsch marks, Italian lire, Netherlands goldens, and Belgian francs. Fluorescent spots also appear on Swiss francs and Canadian dollars. A number of commercial products are manufactured to detect embedded fluorescent denomination-specific security threads.[2]

The Chemical Details

The security thread introduced in the Series 1990 banknotes is a thin metallized polyester strip that is 1.4 to 1.8 mm in width and 10 to 15 μm in thickness.[3] Polyester threads are standard synthetic fibers consisting of large linear (chainlike) or cross-linked (network) polymers formed from a large number of smaller molecules or monomers. The monomers are connected via *ester linkages* as in Fig. 12.6.1. Polyesters most commonly are prepared from equivalent amounts of two different monomers: *glycols* and *dibasic acids*. Glycols are organic compounds containing two hydroxyl groups, –OH, and dibasic acids are organic molecules containing two carboxyl functionalities, –COOH.

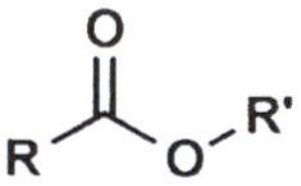

Figure 12.6.1 ▶ Ester linkages that connect monomers to form polymeric molecules known as polyesters.

KEY TERMS:	fluorescence	polyester	ester
	linkage	glycol	dibasic acid

References

[1] "Officials report $100 printing error." National/World News, March 29, 1997, TH On-Line, Telegraph Herald of Dubuque, IA, http://www.thonline.com/th/news/032997/National/52409.htm

[2] "Ultraviolet Applications." UVP Products, Inc., Upland, CA, http://www.uvp.com/html/bulletins.html

[3] "Description and Assessment of Deterrent Features," in *Counterfeit Deterrent Features for the Next-Generation Currency Design*, National Materials Advisory Board (Washington, DC: National Academy Press, 1993), Chap. 4.

Related Web Sites

▶ "Know Your Money." United States Secret Service, United States Treasury, http://www.treas.gov/usss/know_your_money.shtml

▶ "US Treasury and Federal Reserve Introduce New $50 Bill." Treasury News, June 12, 1997, http://www.treas.gov/press/releases/archives/1997.html

▶ "Features of the New Twenty." Bureau of Engraving and Printing, http://www.bep.treas.gov/section.cfm/4/30/57

▶ "Security Features of the Lats Banknotes." Latvijas Banka, http://www.bank.lv/naudas/English/index_security.html

▶ "Your Online Guide to Singapore Currencies: Telling Genuine from Fake Notes: Features Recognizable under Ultraviolet Light." Board of Commissioners of Currency, Singapore, http://www.bccs-sin.com/telling.html

▶ "Counterfeit Money Detector." Trade2000 (USA) Inc., http://www.trade2000.com/intro.htm and http://www.trade2000.com/detector3.htm

▶ "Money Hot off the Press." Bureau of Engraving and Printing, United States Treasury, a 30-second video capsule of how currency is printed, http://www.treas.gov/bep/money.mpg

▶ "The Making of Money." Bureau of Engraving and Printing, United States Treasury, a 39-second video with sound on the currency-making process, http://www.treas.gov/bep/making.mpg

▶ Phosphors for Security, Tagging and UV/IR Detection, Phosphor Technology,
   http://www.phosphor.demon.co.uk/iruv.htm

▶ "Physical Sciences and Technology. New Greenbacks: How to Make a Buck — Literally."
   Richard Lipkin, http://www.sciencenews.org/sn_edpik/ps_4.htm

12.7 Why Do U.S. Bills Shift in Color When Viewed from Different Angles?

After nearly four generations, the currency of the United States has
undergone a noticeable change in appearance using several technological advances. In particular, complex and innovative chemistry has
been incorporated in the design of the ink used to denote the denomination in the lower right-hand corner. Is it an optical illusion, or is the
color of the ink shifting from green to black as you tilt the bill?

The Chemical Basics

With advancements in the technologies of color copiers, scanners, and printers,
maintaining the security of U.S. currency is increasingly difficult. Since 1990 the
U.S. Treasury has added a number of features to new currency as security measures against counterfeiting. One of the new design features for series 1996 notes
is the use of *optically variable ink* on the number in the lower right-hand corner
of the bill. As one tilts the bill in light, the use of a color-shifting ink becomes
apparent. When viewed straight-on, the numerals appear green, but when viewed
at an angle the numbers look black.

The technology was developed by chemist Roger Phillips, currently product
technology manager and manager of intellectual properties at Flex Products in
Santa Rosa in Sonoma County, CA. Phillips and two colleagues, Pat Higgins and
Peter Berning at Optical Coating, originally designed an optically variable foil
that would be applied to U.S. currency as an anticounterfeiting measure. The
Federal Reserve initially liked the idea and spent $17 million to develop the prod-
uct, but by 1985 federal officials decided to abandon the new technology because
the process would require expenditures for new machinery. With the company
facing massive layoffs, Phillips came up with the idea to translate the technology
into a printing ink that would not require costly new equipment.

The United States is not the only government using optically variable ink as
a security feature. Over 40 countries use the pigment technology. On the 500-
lats note issued by the Bank of Latvia (Latvijas Banka), optically variable ink
creates the effect of changing colors for the nominal value printed in the left-
hand corner of the obverse side of the bill. Security measures involving optically
variable ink have also been taken by the Central Bank of Ireland for its bank notes.
The United Kingdom also recently used optically variable ink on first-class rate
stamps reissued in gold to commemorate the 50th anniversary of the marriage of

Her Majesty the Queen to Prince Philip. The Queen's head changes from gold to green with a change in viewing angle. Roger Phillips has also extended the technology of optically variable ink to new areas, for example, revolutionizing the automotive industry. In 1993 he developed a durable automobile paint that changed colors when viewed from different angles.

The Chemical Details

The process for developing optically variable ink involves the layering of several extremely thin metal-containing pigment coatings of precise thickness, followed by grinding of the coating into tiny platelets or flakes. The flakes are typically 1 μm thick and 2 to 20 μm in diameter with an average aspect ratio (i.e., ratio of width to height) of 10 to 1.[1] These color-shifting thin-film flakes are then suspended in a mixture of regular ink, and the ink is then applied to a surface. The high-aspect ratio helps align the flakes parallel to the surface of the ink. As light shines on the flakes, light is reflected. Because of the random positionings of the metallic platelets, certain wavelengths of light are selectively reinforced ("constructive interference"), while other wavelengths are canceled ("destructive interference"). This phenomenon, known as *color diffraction*, creates the appearance of color through reflection (see Fig. 12.7.1).

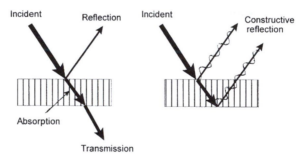

Figure 12.7.1 ▶ The phenomena of light reflection, absorption, and interference to create the appearance of color. Color arises as certain wavelengths of reflected light are selectively reinforced through constructive interference, while other wavelengths of light are canceled through destructive interference.

The particular colors that are observed at different angles will depend critically on the thickness of the thin film coating. Precision instrumentation is required to carefully control film thickness during production. The magnitude of the optical effect depends on the density of flakes in the ink, while the quality of the optical effect depends on the precise orientation or alignment of these flakes with respect to the paper surface.

What are the typical materials used to create the color-shifting flakes? A symmetrical layering pattern of absorber/dielectric/reflector/dielectric/absorber is employed, since the flakes can be oriented either up or down on the ink surface. The role of the *absorber* is to absorb particular wavelengths of light to enhance or modify the observable color change of the ink as the viewing angle is varied. The *dielectric* is a material that does not absorb visible light and also possesses a low refractive index. The interference colors of such materials are highly angle-dependent.[2] The reflector is the material that produces the unique optical effect through the reflection of light. Typical materials used in the multilayer structure are chromium as the absorber, magnesium fluoride or silicon dioxide as the dielectric, and aluminum as the reflector: $Cr/MgF_2/Al/MgF_2/Cr$ and $Cr/SiO_2/Al/SiO_2/Cr$. Layer thicknesses of 50, 4000, 900, 4000, and 50 Å, respectively,[1] are typical. Some additional combinations of materials for optically variable flakes are illustrated in Fig. 12.7.2.

Partially reflective	Al	Cr	MoS_2	Fe_2O_3	Fe_2O_3
Low refraction	SiO_2	MgF_2	SiO_2	SiO_2	SiO_2
Inner reflector	Al	Al	Al	Al	Fe_2O_3
Low refraction	SiO_2	MgF_2	SiO_2	SiO_2	SiO_2
Partially reflective	Al	Al	MoS_2	Fe_2O_3	Fe_2O_3

Figure 12.7.2 ▶   Layering patterns of absorber (partially reflected)/dielectric (low refractive)/reflector (inner reflector)/dielectric/absorber used to create the optical effect of color-shifting ink.

KEY TERMS: reflection   absorber   dielectric   reflector ink

References

[1] "Description and Assessment of Deterrent Features," in *Counterfeit Deterrent Features for the Next-Generation Currency Design*, National Materials Advisory Board (Washington, DC: National Academy Press, 1993), Chap. 4.

[2] "Luster Pigments with Optically Variable Properties." Raimund Schmid, Norbert Mronga, Volker Radtke, Oliver Seeger, BASF Corporation, http://www.coatings.de/articles/schmid/schmid.htm

Related Web Sites

▶ "Know Your Money." U.S. Secret Service, U.S. Treasury, http://www.treas.gov/usss/know_your_money.shtml

▶ "Color-Shifting Ink." Unites States Treasury,
http://www.bep.treas.gov/cd042500/color.html

▶ "Bank Notes and Security Features." Banque de France,
http://www.banque-france.fr/gb/billets/main.htm

▶ "Security Features of the Lats Banknotes." Latvijas Banka,
http://www.bank.lv/naudas/English/index_security.html

▶ "Stories from a Few Who Teach, Innovate, Create." Mary Fricker, The Press Democrat,
Outlook Sonoma County and Its Economy,
http://www.pressdemo.com/outlook97/tech/stories.html

▶ "Money Hot off the Press." Bureau of Engraving and Printing, United States Treasury, a 30-
second video capsule of how currency is printed, http://www.treas.gov/bep/money.mpg

▶ "The Making of Money." Bureau of Engraving and Printing, United States Treasury, a 39-
second video with sound on the currency-making process, http://www.treas.gov/
bep/making.mpg

▶ "Secrets of Making Money." NOVA Online, http://www.pbs.org/wgbh/nova/moolah

12.8 | What Is an Optical Brightener?

"Whiter whites! Brighter brights!" You've heard such claims made by
many manufacturers of laundry detergents. The chemical structure of
the additives called "optical brighteners" provides the essential factors
that make these superior detergents possible.

The Chemical Basics

Natural fibers such as wool and silk and cellulose-containing paper often have a
yellowish tinge. To appear yellow these materials must absorb the complementary
color—a violet to blue hue, i.e., light of wavelength near 400 nm. (The complementary color is the color on the opposite side of a color wheel.) Natural pigments
in these fabrics and in cellulose are responsible for the absorption. Whitening of
fabrics and paper can be accomplished using bleaches, but this treatment often degrades the material. Alternatively, an *optical brightener* may be added to replace
the blue-violet light that is lost (i.e., not transmitted to our eyes but absorbed by
the fiber). An optical brightener is a compound that accomplishes this function by
absorbing ultraviolet light and subsequently emitting ("fluorescing") blue visible
light. The emitted blue light from the optical brightener replaces the blue light
absorbed by the fabric, thereby creating a "complete" white light that contains all
of the frequencies of the color spectrum. The further "brightening" action of an
optical brightener arises when a slight excess of the fluorescent compound is used
to convert even more ultraviolet light into visible light. The amount of optical
brightener used should be carefully controlled, for an excessive level of optical
brightener can lead to a blue cast for the fabric due to the blue emission of light.

The Chemical Details

While many commercial optical brighteners are trade secrets, most of these fluorescent compounds contain one or more ring systems and are derivatives of stilbene (Fig. 12.8.1), coumarin (Fig. 12.8.2), imidazole (Fig. 12.8.3), triazole

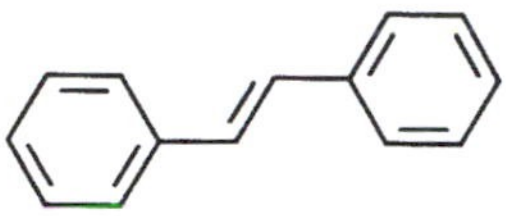

Figure 12.8.1 ► The molecular structure of stilbene.

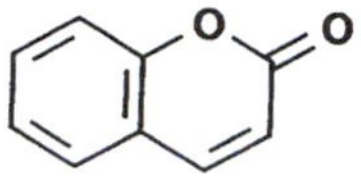

Figure 12.8.2 ► The molecular structure of coumarin.

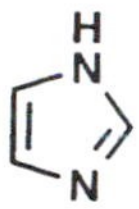

Figure 12.8.3 ► The molecular structure of imidazole.

(Fig. 12.8.4), oxazole (Fig. 12.8.5), and biphenyl (Fig. 12.8.6). The uninterrupted chain of alternating single and double bonds ("conjugated double bonds") in these substances contributes to the absorption of these compounds in the ultraviolet region and their fluorescence in the blue wavelengths of light. One such compound, 7-amino-4-methylcoumarin (Fig. 12.8.7), absorbs near 350 nm and emits at 430 nm. These optical parameters are evident in the normalized absorption and emission spectra[1] of this compound that appear schematically in Fig. 12.8.8. An *absorption spectrum* is a record of how much light of a particular color is absorbed by a substance as a function of color (i.e., wavelength). An emission (or fluorescence) spectrum, a similar concept, measures the intensity of light of a given color emitted by a substance as a function of color (i.e., wavelength). The wavelength of maximum absorption (~350 nm) is in the ultraviolet region of the electromagnetic spectrum. When light principally of this wavelength is absorbed, blue light with a maximum wavelength near 430 nm is emitted.

KEY TERMS: ultraviolet absorption emission fluorescence

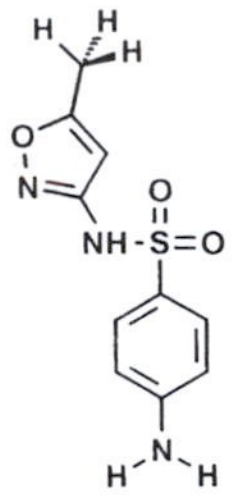

Figure 12.8.4 ▶ The molecular structure of triazole.

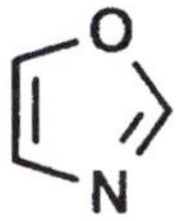

Figure 12.8.5 ▶ The molecular structure of oxazole.

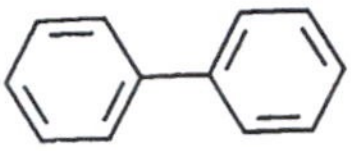

Figure 12.8.6 ▶ The molecular structure of biphenyl.

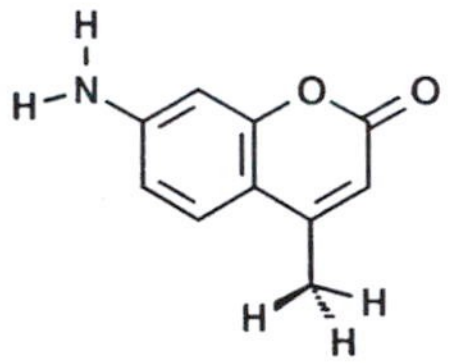

Figure 12.8.7 ▶ The molecular structure of 7-amino-4-methylcoumarin.

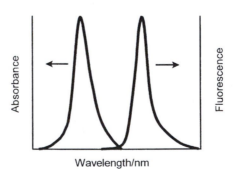

Wavelength/nm

Figure 12.8.8 ▶ A schematic representation of the normalized absorption and emission spectra of 7-amino-4-methylcoumarin.

References

[1] "Spectra: 7-Amino-4-methylcoumarin." Molecular Probes, Inc.,
http://www.probes.com/servlets/spectra?fileid=191ph7

Related Web Sites

► "Structure and Colour in Dyes." Bryan D. Llewellyn,
http://members.pgonline.com/~bryand/StainsFile/dyes/dyecolor.htm

► "The Color Wheel." Painter On-Line, The Maryland Institute, College of Art,
http://www.mica.edu/painter/on_line/colorwhe.htm

12.9   What Is the Difference between a Sunscreen and a Sunblock?

The invisible ultraviolet rays of the sun can cause immediate and long-term skin damage in the form of sunburn, rashes, premature wrinkling, and skin cancer. To avoid overexposure, we are encouraged to apply sunscreens and sunblocks to protect the health of our skin. Chemistry clearly distinguishes between these two formulations, and the chemical structure of these products dictates how well these materials perform.

The Chemical Basics

Sunblocks are opaque substances such as zinc oxide, titanium dioxide, and iron oxide that protect by forming a shield on the skin, which reflects and scatters incident radiation. In essence, sunblocks provide physical protection against sun exposure, including both visible and ultraviolet light. Sunscreens are substances that chemically absorb ultraviolet light in the top layer of the epidermis, protecting the underlying layers.

Visible light ranges in wavelength from 400 to 700 nm. The spectrum of ultraviolet light from the sun ranges in wavelength from 200 to 400 nm and is divided into three broad classifications.[1] UVA rays have the longest wavelength (320–400 nm), are fairly constant year-round, and penetrate deeper into the layers of skin. Shorter UVB rays (290–320 nm) are more intense during the summer months than the longer wavelength UVA radiation. UVB radiation is also stronger at higher altitudes and in areas closer to the equator. UVC radiation, even shorter in wavelength from 200 to 290 nm, is absorbed by the stratospheric ozone layer and does not reach the Earth's surface. (As ozone is depleted in the stratosphere, however, the range of wavelengths of UV light that reaches the Earth's surface will become a greater concern for skin exposure.) The particular chemical structure of a sunscreen determines the wavelengths of ultraviolet light preferentially absorbed by a sunscreen.

The Chemical Details

The sunblocks zinc oxide, titanium dioxide, and iron oxide are inorganic chemicals that are not absorbed into the skin. These substances consist of opaque particles that reflect both visible and ultraviolet light. In addition, zinc oxide blocks virtually the entire UVA and UVB spectrum[1] and thus offers overall protection. The particulate nature of these sunblocks enhances their effectiveness at reflecting sunlight. The smaller the particle size, the greater the surface area available for reflection, and the more effective the sun protection offered by the formulation.[2]

Sunscreens are transparent organic substances that penetrate into the skin and absorb ultraviolet radiation. Common classes of sunscreens include benzophenones, PABA derivatives, cinnamates, salicylates, and dibenzoylmethanes.[3] Benzophenones have a primary protective range in the UVA region and include oxybenzone (Fig. 12.9.1), 270–350 nm; dioxybenzone (Fig. 12.9.2), 206–380 nm;

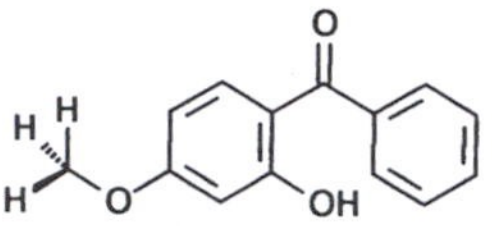

Figure 12.9.1 ▶ The molecular structure of oxybenzone.

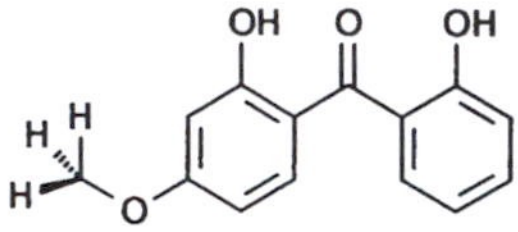

Figure 12.9.2 ▶ The molecular structure of dioxybenzone.

and sulisobenzone (Fig. 12.9.3), 250–380 nm. PABA and PABA esters have

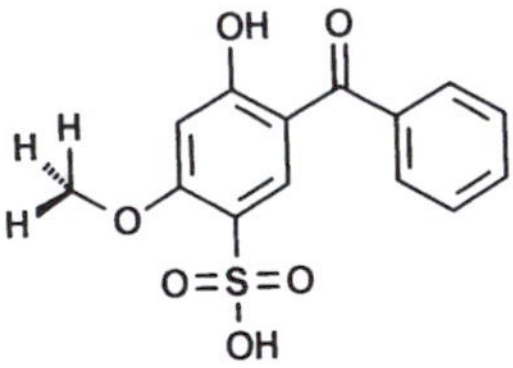

Figure 12.9.3 ▶ The molecular structure of sulisobenzone.

a primary protective range in the UVB range (290–320 nm), including PABA (Fig. 12.9.4), 260–313 nm; Padimate O (Fig. 12.9.5, also known as octyldimethyl PABA), 290–315 nm; Padimate A, 290–315 nm; and glycerol aminobenzoate (Fig. 12.9.6), 260–315 nm. Cinnamates are derivatives of cinnamon, with a primary protective range in the UVB range (290–320 nm). Examples include octyl methoxycinnamate (Fig. 12.9.7), 280–310 nm; and cinoxate (Fig. 12.9.8), 270–

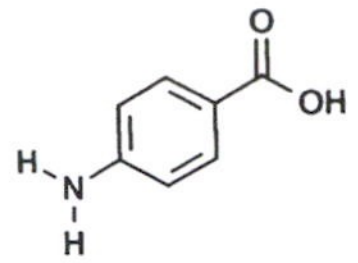

Figure 12.9.4 ▶ The molecular structure of PABA.

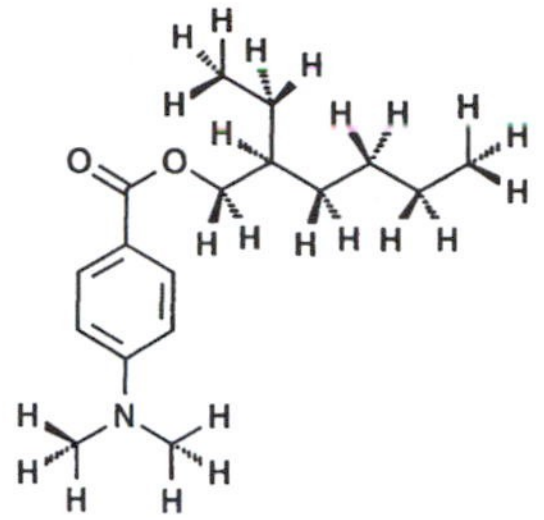

Figure 12.9.5 ▶ The molecular structure of Padimate O (or octyldimethyl PABA).

328 nm. Salicylates, protecting in the UVB range (290–320 nm), include homo-salicylate (Fig. 12.9.9), 290–315 nm; ethylhexyl salicylate (Fig. 12.9.10), 260–310 nm; and triethanolamine salicylate (Fig. 12.9.11), 269–320 nm. Dibenzoyl-methanes serve best as protectors for the UVA range (320–400 nm), offering no protection from UVB. These substances include 4-tert-butyl-4′-methoxydiben-zoylmethane (Fig. 12.9.12), 310–400 nm; and 4-isopropyldibenzoylmethane (Fig. 12.9.13), 310–400 nm. What do all of these molecules have in common that enhances their ability to absorb ultraviolet light?

One common structural element is the presence of an aromatic or benzene ring structure. The aromatic ring is the *chromophore* of sunscreens, that is, the functional group of atoms capable of absorbing ultraviolet light. Benzene, C_6H_6, absorbs in the ultraviolet at 280 nm. The presence of certain substituents on the benzene ring alters the electron distribution in the ring and shifts the absorption wavelength. In addition, the conjugation of double bonds (i.e., the presence of alternating double and single bonds as in C=C–C=C or C=C–C=O) can lead to tremendous shifts in absorption wavelength. In the class of benzophenones, the presence of hydroxyl (–OH) groups on the aromatic rings and the additional

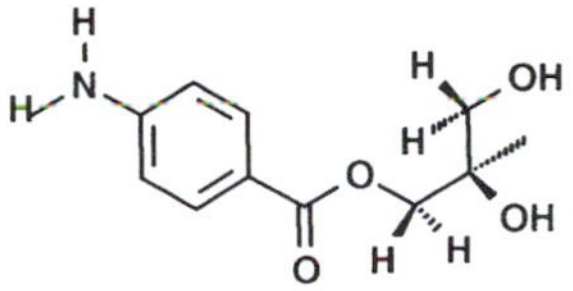

Figure 12.9.6 ▶ The molecular structure of glycerol aminobenzoate.

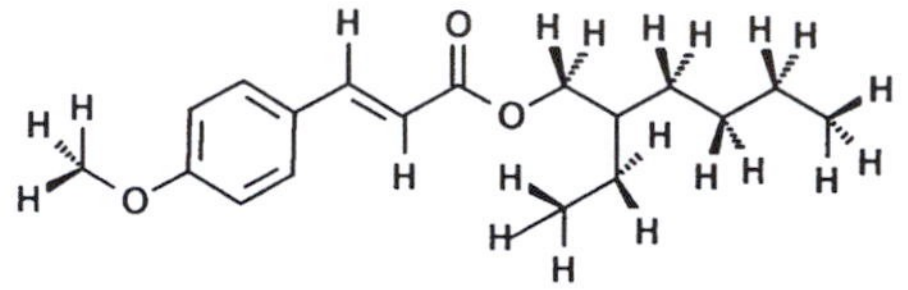

Figure 12.9.7 ▶   The molecular structure of octyl methoxycinnamate.

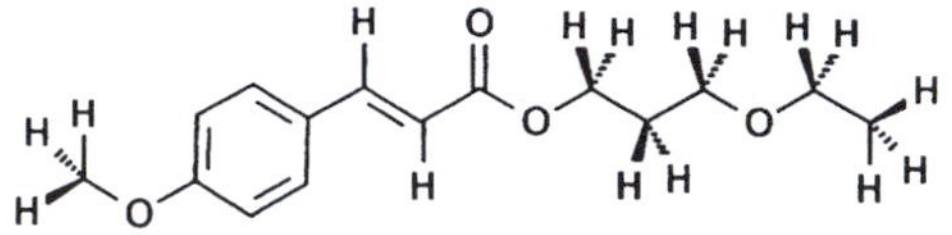

Figure 12.9.8 ▶   The molecular structure of cinoxate.

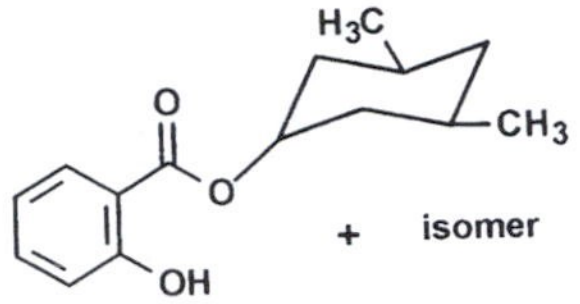

Figure 12.9.9 ▶   The molecular structure of homosalicylate.

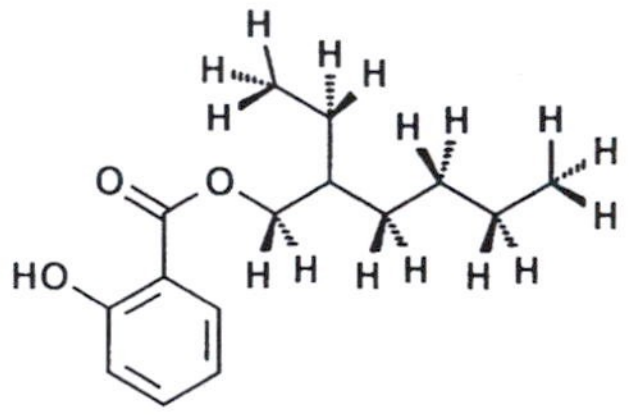

Figure 12.9.10 ▶   The molecular structure of ethylhexyl salicylate.

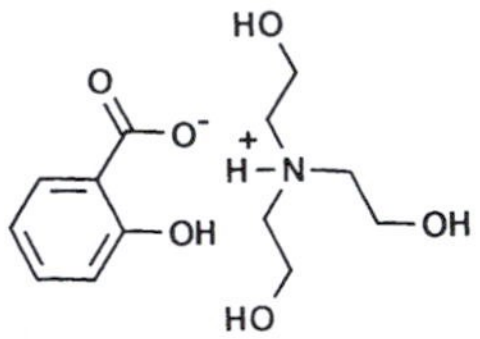

Figure 12.9.11 ▶   The molecular structure of triethanolamine salicylate.

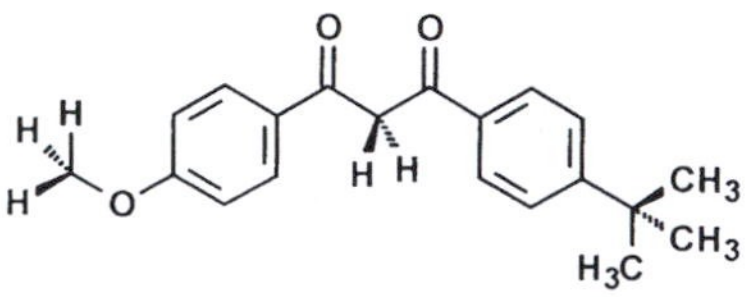

Figure 12.9.12 ▶   The molecular structure of 4-tert-butyl-4′-methoxydibenzoylmethane.

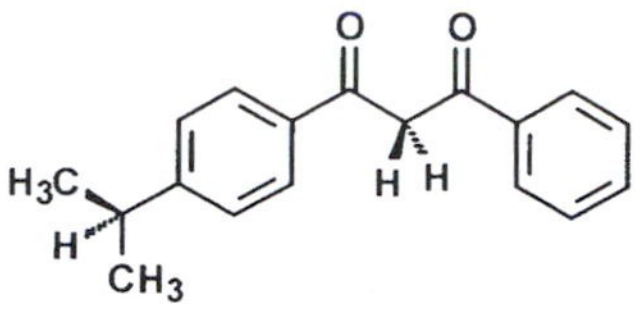

Figure 12.9.13 ▶   The molecular structure of 4-isopropyldibenzoylmethane.

conjugation shift the absorption to longer wavelengths, particularly in the UVB range. Amino substituents ($-NH_2$) and carboxyl groups ($-COOH$) also increase the wavelength of absorption. Sunscreens thus have a chemical structure that matches their function — the absorption of ultraviolet light.

KEY TERMS: **ultraviolet**   **chromophore**   **conjugation**

References

[1] "The Protective Role of Zinc Oxide: Sunscreens." Mark Mitchnick, sunSmart Inc., http://www.iza.com/zhe_org/Articles/Art-09.htm

[2] "TiO$_2$sperse? Ultra: Product Profile, Collaborative Laboratories, http://www.collabo.com/tioultra.htm

[3] "Sunscreens." http://www.geocities.com/HotSprings/4809/sunscr.htm# ACTIVE IN-GREDIENTS IN SUNSCREENS

Other Questions to Consider

1.2   Why does a kitchen gas burner glow yellow when a pot of boiling water overflows? *See* p. 2.

5.4   Why is hydrogen peroxide kept in dark plastic bottles? *See* p. 40.

5.20   Why does chlorine in swimming pools work best at night? *See* p. 70.

Chapter 13 ▶

Connections to Organic Chemistry

13.1 How Do Sutures Dissolve?

History tells us that catgut suture had its origin around AD 150 in the time of the Greek physician Galen, who built his reputation by treating wounded gladiators.[1,2] What chemical materials do modern surgeons use for sutures to ensure biocompatibility, tensile strength, and dissolution by the body's natural action?

The Chemical Basics

Have you ever wondered how sutures could possibly hold a wound together long enough to promote healing and manage to dissolve over time? And what about the added need for flexibility so that the surgeon can tie a knot that holds? Synthetic biodegradable polymers have the desirable mechanical and chemical properties to perform all of these functions. Some biodegradable polymers degrade by hydrolysis, i.e., by a reaction with water; others degrade enzymatically, i.e., via a reaction with an enzyme in the body. The products of the degradation process are harmless to the body.

The Chemical Details

Common biodegradable polymers for medical devices are constructed from synthetic linear aliphatic polyesters. One material commonly used for internal sutures is poly(glycolic acid) (PGA). PGA is synthesized from the dimer of glycolic acid (Fig. 13.1.1).[1]

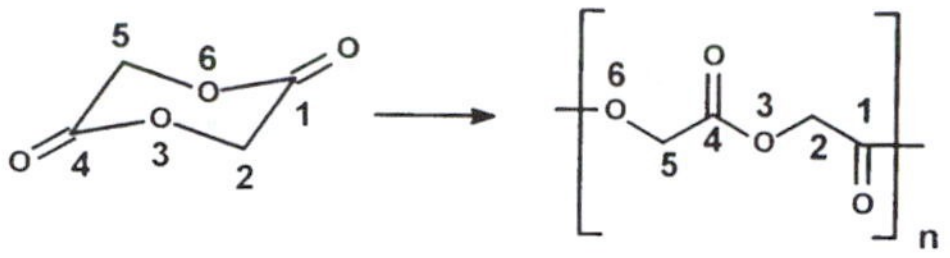

Figure 13.1.1 ▶ The synthesis of poly(glycolic acid) (PGA) from the dimer of glycolic acid.

PGA degrades by hydrolysis to produce carbon dioxide and glycolic acid (Fig. 13.1.2), which is either excreted or enzymatically converted to other metabo-

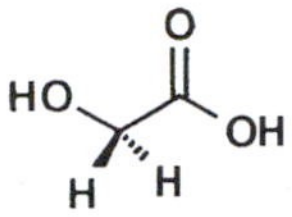

Figure 13.1.2 ▶ The molecular structure of glycolic acid.

lized species. Lactic acid (Fig. 13.1.3) has also been polymerized into poly(lactic

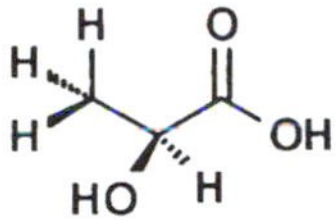

Figure 13.1.3 ▶ The molecular structure of lactic acid.

acid) (PLA) and developed into commercial sutures. PLA is prepared from the cyclic diester of lactic acid (lactide) by ring-opening polymerization (Fig. 13.1.4).[1]

The lactic acid generated by the hydrolytic degradation of PLA becomes incorporated into one of the normal metabolic cycles of the body and is excreted as carbon dioxide and water. Copolymers of glycolic acid and lactic acid are also commonly used as biodegradable sutures to balance the greater strength of PGA and the slower degradation of PLA.

KEY TERMS: polymers   hydrolysis

References

[1] "Bioadsorbable Polymers." BMEn 5001, *An Introduction to Biomaterials*, William B. Gleason, University of Minnesota, Site Index,
http://www.courses.ahc.umn.edu/medical-school/BMEn/5001/

[2] "Biomedical Applications of Textiles: Sutures." Alice Baker, Will Fowler, Sophie Guevel, Allen Smith, North Carolina State University,

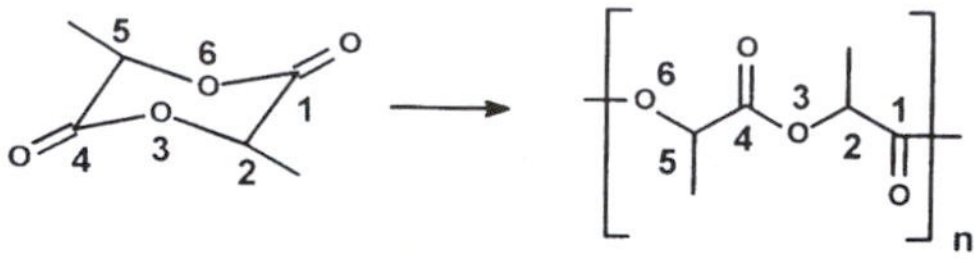

Figure 13.1.4 ▶   The synthesis of poly(lactic acid) (PLA) by a ring-opening polymerization of the cyclic diester of lactic acid (lactide).

http://www.bae.ncsu.edu/bae/research/blanchard/www/465/textbook/
biomaterials/projects/textiles/fowler/project1.html

Related Web Sites

▶ "A Brief History of the Origins of Sutures." The Veterinarian's Sutures Guide, Dr. R.A. Henderson, http://www.vetmed.auburn.edu/~hendera/guide/guide2.htm

▶ "Materials." Dr. David Harrison, School of Education, North East Wales Institute of Higher Education, Wrexham, North Wales, http://www.newi.ac.uk/buckleyc/materials.htm#Polymers and Plastics

▶ Medical Devices, Birmingham Polymers, Inc., http://www.birminghampolymers.com/htdocs/medical_devices.htm

13.2 How Do You Deodorize Skunk Spray on Your Dog?

One pet owner recently prescribed the following home remedy to eliminate skunk odor: "Add 1/4 cup of baking soda and 1 teaspoon of liquid detergent to 1 quart of hydrogen peroxide. Soak a rag with the solution and saturate the affected areas, rubbing it in."[1] What valuable chemistry is working to deodorize your dog?

The Chemical Basics

The odor from skunk spray originates from a substance that belongs to a family of compounds with many odiferous members. This class of substances, known as *thiols* or mercaptans, is responsible for the delectable aroma of freshly brewed coffee (an ingredient called furfurylthiol), the repugnant smell of rotten eggs (a substance known as hydrogen sulfide), and the warning odor of natural gas for leak detection (an odorant called 2-methyl-2-propanethiol). The sensitivity of the human nose to these compounds, often at levels as low as 20 ppb, contributes to their reputation as pungent odors. By a chemical conversion of a thiol to another class of compound, the disagreeable skunk scent can be removed.

The Chemical Details

Skunks deter predators by release of a liquid spray containing seven major volatile components[2] classified as thiols (compounds containing the –SH functional group) and acetate derivatives of thiols (characterized by the –SC(O)CH$_3$ functionality). In particular, two of the more odiferous components responsible for the strongly repellent odor of the skunk's secretion are 2-butene-1-thiol (Fig. 13.2.1) and 3-

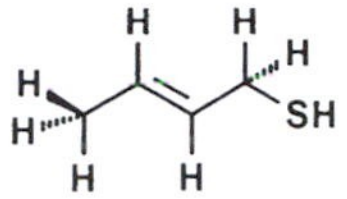

Figure 13.2.1 ▶ 2-Butene-1-thiol, an odiferous component of a skunk's secretion.

methylbutane-1-thiol (Fig 13.2.2.).[2]

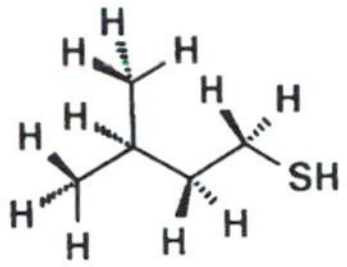

Figure 13.2.2 ▶ 3-Methylbutane-1-thiol, an odiferous component of a skunk's secretion.

To deodorize a pet sprayed by a skunk, the noxious compound must be converted to an odorless one. Using an alkaline solution of 3% hydrogen peroxide and sodium bicarbonate, the thiol compound (denoted R–SH) can be oxidized to a disulfide compound,

$$2\,RSH \rightarrow RS\text{–}SR + 2\,H^+ + 2\,e^-.$$

or in the presence of an alkaline substance such as detergent

$$2\,RSH + 2\,OH^- \rightarrow RS\text{–}SR + 2\,H_2O + 2\,e^-.$$

Here hydrogen peroxide acts as the oxidizing agent:

$$H_2O_2\,(aq) + 2\,e^- \rightarrow 2\,OH^-\,(aq).$$

Thus, the overall reaction is written

$$\underset{\text{odoriferous}}{2\,RSH} + H_2O_2\,(aq) \rightarrow \underset{\text{odorless}}{RS\text{–}SR} + 2\,H_2O.$$

KEY TERMS: **thiol**   **oxidation–reduction**

References

[1] "Skunk Odor Removal." Cooperative Extension in Lancaster County, Penn State, http://lancaster.extension.psu.edu/Family/newsletters/HomeLlfeJulAug.htm #Skunk_Odor

[2] "Chemistry of Skunk Spray." Professor William F. Wood, Department of Chemistry, Humboldt State Univeristy, http://sorrel.humboldt.edu/~wfw2/chemofskunkspray.html

Related Web Sites

► "Deodorize Skunk Spray: Neutralize Skunk Odor." Professor William F. Wood, Department of Chemistry, Humboldt State University, Arcata, CA, http://www.humboldt.edu/ ~wfw2/deodorize.shtml

13.3 | How Do Forensic Chemists Use Visible Stains to Trap Thieves?

Thief detection powder is designed for thief detection and the identification of stolen or altered items. Once a marked article is touched, the powder creates a highly colored visible stain on the skin. What chemistry do forensic specialists apply to investigate crimes using detection powders?

The Chemical Basics

Increasingly law enforcement agencies, banks, businesses, corporate offices, and concerned citizens are using detection powders to trap thieves. Thief detection powders are designed to be applied to the surface of an object likely to be stolen such as cashboxes, cash, or sensitive documents. Alternatively, the powder may be applied to doorknobs or tools to detect entry or use. When the powder comes in contact with the skin, a reagent in the powder reacts with the body's amino acids to create a highly visible long-lasting purple-colored stain. While the stain will not immediately wash off the skin, the color will gradually rub off after about a day.

The Chemical Details

Detection powders and fingerprint development kits commonly contain cream- or yellow-colored ninhydrin crystals or a solution of dissolved ninhydrin. Ninhydrin (also known as 1,2,3-indantrione, monohydrate; 2,2-dihydroxy-1,3-indandione; triketohydrindene, monohydrate; and triketohydrinden hydrate) has the structure presented in Fig. 13.3.1. Ninhydrin will react with a free α-amino group, $-NH_2$. This group is contained in all amino acids, and analysis with ninhydrin is often performed to verify the presence of amino acids. When α-amino acids (i.e., amino acids with the structure $NH_2-CHR-COOH$) react with ninhydrin, a characteris-

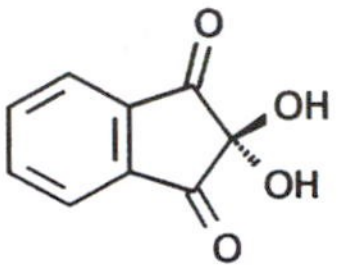

Figure 13.3.1 ▶ The molecular structure of ninhydrin.

tic deep blue or purple color of reduced ninhydrin is observed. The reactions involved in this oxidation–reduction process are shown in Fig. 13.3.2.[1]

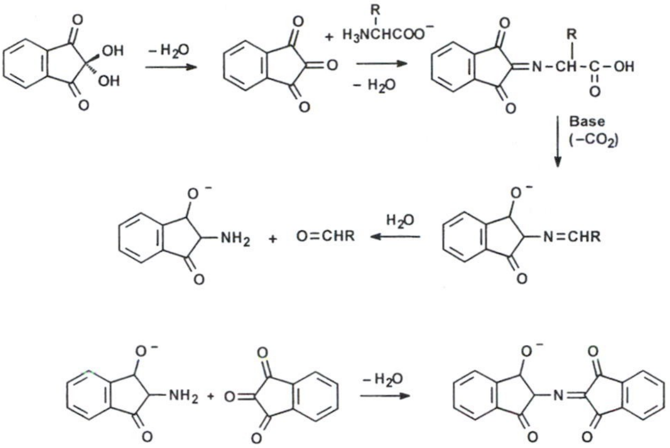

Figure 13.3.2 ▶ The sequence of reactions that reduce ninhydrin to produce a blue color.

KEY TERMS: *α*-amino acid **amine group**

References

[1] "Ninhydrin Reaction with Amino Acids, Biochemistry Survey Lecture." D. S. Moore, Department of Chemistry, Howard University,
http://www.chem.howard.edu/~dmoore/biochemlecs/Lec-11/Ninhydrin.html

Related Web Sites

▶ "The Art and Science of Criminal Investigation: Ninhydrin Processing." *Crimes and Clues*, Pat A. Wertheim, http://www.crimeandclues.com/ninhydrin.htm

▶ "Visible Stain Thief Detection Powder." The Spy Store,
http://www.spy-store.com/Theftpowders1.html

13.4	**Why Is White Willow Bark Known as "Nature's Aspirin"?**

The Royal Society received a communication from the Reverend Edmond Stone of Chipping Norton in Oxfordshire on June 2,1763.[1] The opening lines are as follows:

> Among the many useful discoveries which this age has made, there are few which better deserve the attention of the public than what I am going to lay before your Lordship. There is a bark of an English tree, which I have found by experience to be a powerful astringent and very efficacious in curing agues and intermittent disorders.

The Chemical Basics

Aspirin, or acetylsalicylic acid, is the world's most widely used remedy for reducing pain and lowering fever. Although aspirin's first description in the medical literature in 1899 was for the treatment of rheumatic fever, a simpler derivative of aspirin known as *salicylic acid* had been described earlier (1876) as being effective for controlling fever and treating gout and arthritis. Still earlier (1763), homeopathic medical practitioners reported that the chewing of white willow (*Salix*) bark was effective in treating malaria. Only later was it determined that extraction of willow bark yields salicin, a compound that can be hydrolyzed and oxidized to salicylic acid.

The Chemical Details

Acetylsalicylic acid is an example of an ester, a class of organic compounds with the general formula RCOOR′. Esters contain an "ester linkage" consisting of a carbonyl group to which an alkyl group (R) and an alkoxy group (OR′) are bonded (Fig. 13.4.1). Under either acidic or basic conditions or in the presence

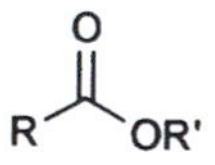

$$\underset{R}{\overset{O}{\underset{\Large\|}{}}}\overset{}{\diagup}\overset{}{\diagdown}OR'$$

Figure 13.4.1 ▶ A general ester structure.

of certain enzymes, esters undergo a *hydrolysis* reaction to break the ester linkage and decompose the ester structure into two smaller compounds, a carboxylic acid and an alcohol:

$$\text{ester} + \text{water} \xrightarrow{\text{H}^+ \text{ or OH}^- \text{ or enzyme}} \text{carboxylic acid} + \text{alcohol.}$$

Salicylic acid (Fig. 13.4.2) is an example of a carboxylic acid with the general formula R–COOH. In this class of organic compounds the carbonyl functionality

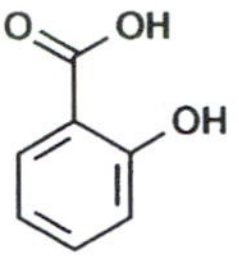

Figure 13.4.2 ▶ The molecular structure of salicylic acid.

(C=O) is linked to both an alkyl group (or hydrogen atom) and a hydroxyl group. The medicinal activity of salicylic acid suggests that it is this phenolic precursor (phenol, Fig. 13.4.3), not acetylsalicylic acid (Fig. 13.4.4), that is the prime pain

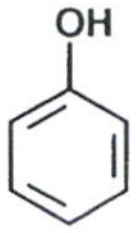

Figure 13.4.3 ▶ The molecular structure of phenol.

reliever.

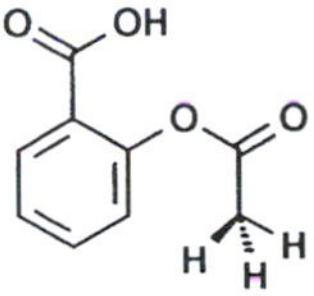

Figure 13.4.4 ▶ The molecular structure of aspirin, acetylsalicylic acid.

Salicin (Fig. 13.4.5), on the other hand, belongs to the class of organic compounds known as *ethers*. Ethers contain an oxygen atom (the "ether linkage") to which two alkyl groups (or their derivatives) are bonded: R–O–R'. Ethers are synthesized via a condensation reaction of two alcohols (generating the ether and a water molecule). The reverse reaction is possible also; ethers can undergo decomposition reactions to yield the two constituent alcohols. Under oxidation conditions (e.g., in the presence of air or oxygen or another oxidizing agent), a *primary alcohol* created during the ether decomposition can undergo oxidation to an aldehyde and further oxidation to a carboxylic acid. A primary alcohol (1°) contains a hydroxyl group (–OH) bonded to a carbon atom to which only one other carbon atom is attached: RCH_2OH. A secondary alcohol (2°) with the general formula $RCHR'OH$ and a tertiary alcohol (3°) with the general formula $RCR'R''OH$ each contain hydroxyl groups attached to a carbon atom with two and three attached carbon atoms, respectively.

Decomposition of an ether:

$$\text{ether} + \text{water} \xrightarrow{H_2SO_4} \text{alcohol \#1} + \text{alcohol \#2}$$

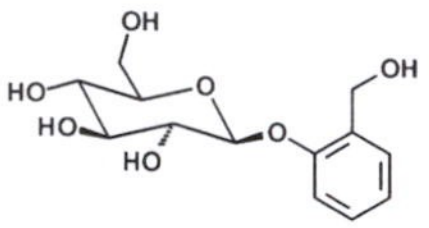

Figure 13.4.5 ▶   The molecular structure of salicin.

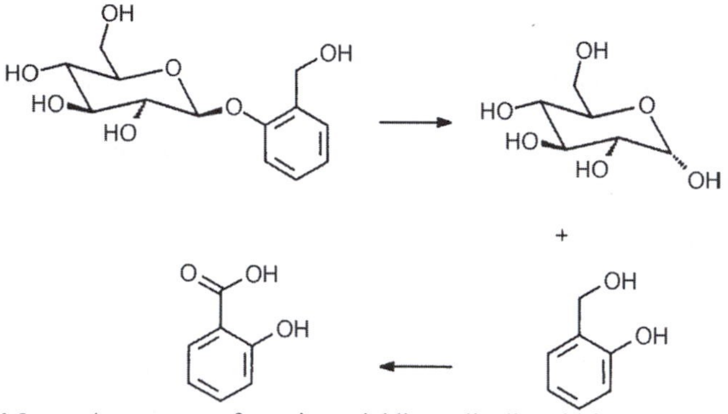

Figure 13.4.6 ▶   A sequence of reactions yielding salicylic acid from the natural product salcin from white willow bark.

Oxidation of a primary alcohol:

$1°$ alcohol $\xrightarrow{O_2,\ K_2Cr_2O_7,\ \text{or KMnO}_4}$ aldehyde $\xrightarrow{O_2,\ K_2Cr_2O_7,\ \text{or KMnO}_4}$ carboxylic acid

Decomposition of salicin (2-(hydroxymethyl)phenyl-β-D-glucopyranoside) yields two alcohols: glucose (a compound with numerous $1°$, $2°$, and a $3°$ hydroxyl functionalities, Fig. 13.4.6) and 1–OH-2-CH_2OH-benzene (a $1°$ alcohol, Fig. 13.4.7). Further oxidation of 1–OH-2-CH_2OH-benzene leads to the formation of salicylic acid:

salicin $\xrightarrow{H_2O,\ H^+}$ two alcohols $\xrightarrow{O_2\ \text{or enzyme}}$ salicylic acid

ether $\longrightarrow$ glucose + 1-OH-2-CH_2OH-benzene($1°$ alcohol) $\longrightarrow$ salicylic acid.

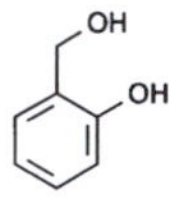

Figure 13.4.7 ▶   The molecular structure of 1-OH-2-CH_2OH-benzene.

KEY TERMS: ester   hydrolysis   ether   alcohol

References

[1] G. Weissmann, "NSAIDs: Aspirin and Aspirin-Like Drugs," in *Cecil Textbook of Medicine*, 19th ed., ed. J. B. Wyngarden, L. H. Smith, J. C. Bennett (Philadelphia: W. B. Saunders, 1988), 114.

Related Web Sites

▶ "Aspirin: A New Look at an Old Drug." Ken Flieger, Federal Consumer Information Center of the U.S. General Services Administration, Pueblo, CO, http://www.pueblo.gsa.gov/press/nfcpubs/aspirin.txt

▶ "White Willow Bark." Healthy Wave, 1999, http://www.healthywave.com/ingredients/whitewillowbark.html

▶ "The Patient on Low-Dose Aspirin For Neuraxial Anesthesia Part I." McMahonMed.com, http://www.mcmahonmed.com/cme/an/cme176/05lesson.htm

▶ "A Miracle Drug." Sophie Jourdier, ChemBytes e-zine, chemsoc, http://www.chemsoc.org/chembytes/ezine/1999/jourdier.htm

13.5 | Why Does Balsamic Vinegar Have a Sweet Taste?

The word *vinegar* is derived from the French *vin aigre* meaning "sour wine." Why, then, does balsamic vinegar have a sweet taste?

The Chemical Basics

Vinegar results from the conversion of alcoholic solutions (wine, beer, or cider) into acetic acid by the action of Acetobacter aceti bacteria. A rich variety of vinegars exist. For example, apple cider vinegar is made from fermented apple cider, while white vinegar is produced from a grain-alcohol mixture. Red and white wine vinegars are derived, as expected, from red and white grape wine, malt vinegar is obtained from malted barley, and rice and cane vinegar from the grain rice and sugar cane, respectively. Balsamic vinegar is unique among traditional Western vinegars, because it does not start out as fermented wine but rather as unfermented Trebbiano white grape juice or must. This mature grape delivers a sweet juice. The grape juice is boiled down to almost a syrup and then proceeds through a natural fermentation process in wooden casks to produce alcohol. A second fermentation with the aid of acetobacter bacteria in the air creates acetic acid. This vinegar is then filtered into wooden casks and left to mature for anywhere from 10–30 years. The complex balance of sweetness from residual sugar in the Trebbiano grape and tartness from the acetic acid produced in the fermentation process gives balsamic vinegar its prized taste.

The Chemical Details

Acetobacter bacteria oxidatively convert wine to vinegar through an aerobic fermentation of ethanol (a primary alcohol) into acetic acid (a carboxylic acid):

$$\text{ethanol} + \text{oxygen} \longrightarrow \text{acetic acid} + \text{water}$$

$$CH_3CH_2OH + O_2 \longrightarrow CH_3COOH + H_2O.$$

In an fermentation process of a solution containing sucrose, the enzyme invertase, present in yeast, acts as a catalyst to convert sucrose into a 1:1 mixture of glucose and fructose. Thus, sucrose is a disaccharide that hydrolyzes in the presence of certain bacteria to yield glucose and fructose. The ether linkage in sucrose is broken to yield two alcohols:

$$\underset{\text{Sucrose}}{C_{12}H_{22}O_{11}} + H_2O \overset{\text{invertase}}{\longrightarrow} \underset{\text{Glucose}}{C_6H_{12}O_6} + \underset{\text{Fructose}}{C_6H_{12}O_6}.$$

The glucose and fructose formed are then converted into ethanol and carbon dioxide by another enzyme, zymase, which is also present in yeast:

$$C_6H_{12}O_6 \longrightarrow 3\ CH_3CH_2OH + 2\ CO_2.$$

KEY TERMS: **carboxylic acid**   **alcohol**   **oxidation**   **disaccharide**

13.6 How Do "Sniffing Dogs" Detect Narcotics or Explosives?

Dogs that can detect narcotics or explosives are in high demand by law enforcement agencies. What chemistry do these dogs know that enables them to reliably aid police in apprehending criminals?

The Chemical Basics

Dogs (in fact, most animals) have a keen sense of smell that enables them to detect certain substances on a parts-per-billion (ppb) level. Are dogs focusing on a specific drug or explosive when they sniff out illegal substances? While many dogs can detect the smell of various targeted drugs, generally they are alerted to various drug-decomposition products or solvents associated with the drug manufacture. These substances generally have vapor pressures higher than those of the drugs themselves and are more readily detected.

The Chemical Details

For example, dogs trained to detect cocaine often detect methyl benzoate,[1] a product of the partial decomposition of cocaine when it is exposed to humid air.[2] Note that cocaine (Fig. 13.6.1) has two ester linkages; these ester linkages are

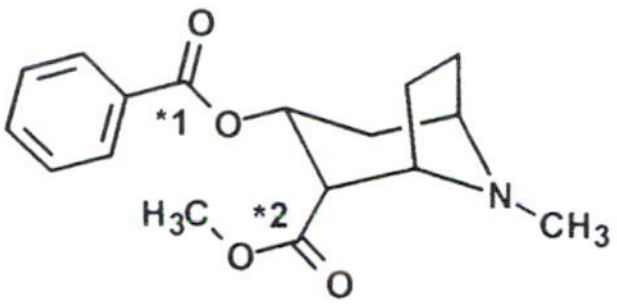

Figure 13.6.1 ▶   The molecular structure of cocaine.

denoted with an asterisk at the carbonyl carbon. Hydrolysis of an ester produces a carboxylic acid and an alcohol. Hydrolysis of cocaine can lead to methylbenzoic acid as one decomposition product at the carbonyl carbon marked *1; methanol can result from the ester linkage decomposition at *2. These two by-products can combine to form a new ester, methyl benzoate.

Heroin is often detected by the presence of acetic acid vapors since acetic acid is a normal decomposition product of heroin.[2] In addition, heroin produced in illegal drug laboratories by treating morphine (see Fig. 13.6.2) with acetic anhy-

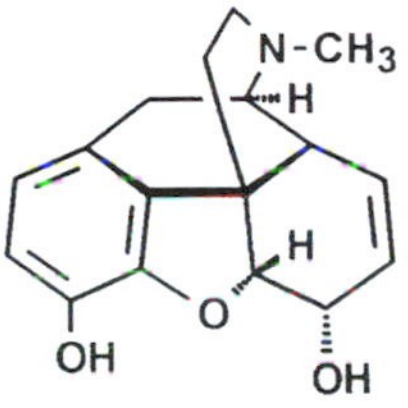

Figure 13.6.2 ▶   The molecular structure of morphine.

dride often yields acetic acid as a by-product. Note that the hydroxyl groups in morphine are acetylated in forming heroin (Fig. 13.6.3) (i.e., the –OH function-

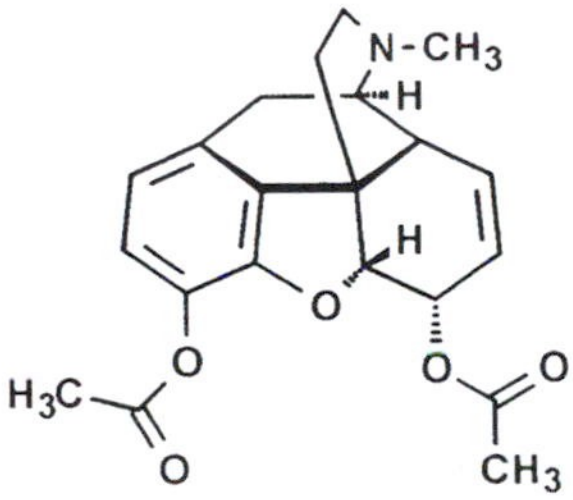

Figure 13.6.3 ▶   The molecular structure of heroin.

ality is converted to the acetyl group –OCOCH$_3$). This reaction is an *esterification* — a reaction of an alcohol (the –OH functionality) and a carboxylic acid (i.e., acetic acid) to produce an ester and water.

Volatile solvents such as toluene, acetone, benzene, and various amyl- and butyl-alcohols are often present in trace amounts as a consequence of solvent washes during the synthesis of the illegal drugs. The sensitivity of canines to these solvents is often the means by which narcotics are detected.

As a further example of a dog's ability to discriminate certain solvents, it is believed that dogs searching for the explosive C-4 as a target substance may be detecting the presence of cyclohexanone rather that the explosive itself. For example, cyclohexanone is a volatile solvent used in the purification process for C-4.[3]

KEY TERMS: ester   **hydrolysis**   **volatility**

References

[1] "Scent as Forensic Evidence and Its Relationship to the Law Enforcement Canine," Charles Mesloh, http://www.uspcak9.com/training/forensicScent.pdf

[2] "Chemistry- and Biology-based Technologies for Contraband." *SPIE Proceedings*, Vol. 2937, http://www.spie.org/web/abstracts/2900/2937.html

[3] "Cargo Inspection Technologies." *SPIE Proceedings*, Vol. 2276, 1994, http://www.spie.org/web/abstracts/2200/2276.html

Related Web Sites

▶ "Dogs Detecting Drugs." Bulletin on Narcotics — 1976 Issue 3–004, United Nations Office for Drug Control and Crime Prevention, http://www.undcp.org/bulletin/bulletin_1976-01-01_3_page005. html

13.7 Why Should You Avoid Acidic Foods When Taking Penicillin?

"Take this medication on an empty stomach 1 hour before meals or 2 hours after meals and avoid the following foods during the time this medication is prescribed: tomatoes, pickles, citrus fruits and juices, coffee, colas, vinegar, and wine." Why does food affect a prescription drug like penicillin?

The Chemical Basics

Penicillin is quite unstable under acid conditions. The penicillin molecule is subject to several hydrolytic reactions, all yielding therapeutically inactive compounds. The stomach is highly acidic (with a pH about 1–3) both to destroy bacteria that are ingested and to activate the enzyme pepsinogen that initiates the digestion of proteins. Enhancing the acidity of the stomach contents due to the ingestion of acidic foods increases the likelihood of producing nontherapeutic forms of penicillin.

The Chemical Details

One of the reactions that occurs under acidic conditions involves the hydrolysis of the amide linkages in penicillin. The structure of penicillin (Fig. 13.7.1) indicates

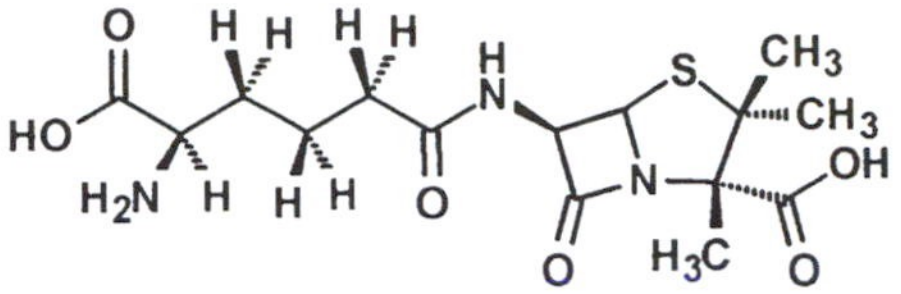

Figure 13.7.1 ▶ The molecular structure of penicillin.

two amide bonds. Amides react in the presence of water and hydrogen ions to yield a carboxylic acid and an amine:

$$\text{Amide} + \text{water} \xrightarrow{\text{acid}} \text{carboxylic acid} + \text{amine}.$$

The β-lactam ring (the four-membered ring) of penicillin has a significant level of ring strain and is thus more susceptible to nucleophilic attack than a generic amide (i.e., the amide functionality between the aromatic and β-lactam rings). Ring strain is reduced when the nucleophile attacks the carbonyl group. Recall that a bond angle in a four-membered ring is constrained to be roughly $90°$, which can result in significant angle strain, thus destabilizing the molecule.

KEY TERMS: amide   hydrolysis   carboxylic acid   amine

Related Web Sites

▶ "Penicillin Derivatives." Gordon L. Coppoc, School of Veterinary Medicine, Purdue University, http://www.vet.purdue.edu/depts/bms/courses/chmrx/penems.htm

Other Questions to Consider

7.7   How does a bullet-proof vest work? How is it made? *See* p. 92.

CLASSES OF ORGANIC COMPOUNDS

13.8   Why Do We Detect Odor in an "Odorless Gas" Leak?

Natural gas: a clean, reliable, and energy-efficient fuel; an odorless, colorless, combustible fuel. These attributes accurately characterize the form of energy known as natural gas. Nevertheless, we are constantly advised to contact the gas service company whenever we detect the smell of gas as a precautionary measure to locate leaks. An examination of the chemical composition of natural gas can explain this paradox.

The Chemical Basics

Natural gas is actually a mixture of several combustible gases found in deposits in the Earth's crust. Natural gas is composed of a mixture of low-molecular-weight *hydrocarbons* (compounds composed of the two elements carbon and hydrogen), primarily methane (CH_4) and ethane (C_2H_6). These gases are indeed odorless and also colorless. However, other more pungent gaseous constituents may be present in natural gas. Some of these gases are inherently present in gas reservoirs as a consequence of the decay of the organic matter in the fossils of plants and animals that produced natural gas millions of years ago. One such gas, hydrogen sulfide — a member of the *mercaptan* or *thiol* family — has the odor of rotten eggs. Other odorous gases are deliberately added to natural gas by your gas service company as a safety precaution to facilitate the detection of both indoor and outdoor gas leaks.

The Chemical Details

The hydrocarbon mixture in natural gas reservoirs primarily contains two low-molecular-weight gaseous compounds, methane (CH_4) and ethane (C_2H_6), both of which are gaseous under atmospheric conditions. Heavier hydrocarbons such as propane (C_3H_8), butane (C_4H_{10}), pentane (C_5H_{12}), and hexane (C_6H_{14}) are also likely.[1] Below the Earth's surface, the high pressures ensure that even propane and butane (normally gases at atmospheric pressure) exist as liquid species known as liquefied petroleum gas. As these substances rise to the Earth's surface, the lower pressures vaporize these substances to produce petroleum gas.

Other gases are often commonly present in natural gas deposits, including hydrogen, nitrogen, and carbon dioxide. Volcanic activity and the decay of organic matter also produce trace amounts of hydrogen sulfide. Even the noble gases argon and helium are constituents of natural gas, produced from natural radioactive disintegration of radioisotopes of potassium (Ar) and thorium and uranium (He). Of these component gases, the extremely penetrating odor of even a small amount of naturally occurring hydrogen sulfide, H_2S, imparts a detectable odor to natural gas. The human nose is quite sensitive to sulfur-containing compounds in the *thiol* class (containing the –SH functional group) with detection limits as low as 18 ppb.[2] Odorants are also added to natural gas to enhance the detection of leaks. The compounds ethanethiol (Fig. 13.8.1), CH_3CH_2SH, and 2-methyl-2-

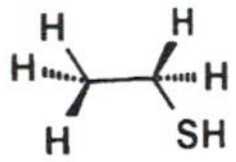

Figure 13.8.1 ▶   Ethanethiol, CH_3CH_2SH, an example of an odorant and warning agent for natural gas.

propanethiol or *tert*-butyl mercaptan (Fig. 13.8.2) are examples of *odorants* and

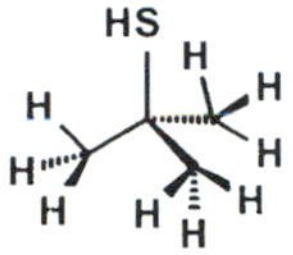

Figure 13.8.2 ▶   2-Methyl-2-propanethiol or *tert*-butyl mercaptan, an example of an odorant and warning agent for natural gas.

warning agents for natural gas. These substances are liquids at atmospheric pressure (*normal boiling points*, i.e., boiling points at 1 atm, of 35–64°C, respectively, for ethanethiol and 2-methyl-2-propanethiol)[3] but can exist as gases under pressure at room temperature.

KEY TERMS: hydrocarbon   thiol

References

[1] "Natural Gas, Manufactured Gas & Liquefied Gas Analysis: Part I — Background." Atlantic Analytical Laboratory, Inc., http://www.test-lab.com/gasone.htm

[2] "Smell-Seeing: A New Approach to Artificial Olfaction." K. S. Suslick, M. E. Kosal, N. A. Rakow, and A. Sen, School of Chemical Sciences, University of Illinois at Urbana-Champaign, http://www.scs.uiuc.edu/suslick/pdf/eurodeur2001.pdf

[3] ChemFinder, Cambridge Soft Corp., http://chemfinder.camsoft.com/

Related Web Sites

▶ "Odor Dispersion: Models and Methods." Richard J. Pope and Phyllis Diosey, New York Water Environment Association Inc., Clearwaters 30 No. 2 (2000), http://www.nywea.org/302140.html

13.9 Why Must Ammonia Cleansers Never Be Mixed with Bleach?

Both ammonia and bleach are useful household products for cleaning stains, sanitizing surfaces, disinfecting, and deodorizing. However, the chemistry of these individual products dictates that you always heed the warning "Caution: Never mix bleach and ammonia cleansers!"

The Chemical Basics

Liquid household bleach is generally a 5% solution of sodium hypochlorite (NaOCl). Ammonia cleansers — including general household cleansers, wax removers, glass and window cleaners, and oven cleaners — are aqueous solutions of 5–10% ammonia, NH_3. Mixing bleach with cleansers containing ammonia leads to the formation of a family of potentially toxic compounds known as *chloramines*. These toxic gases have acrid fumes that can burn mucous membranes. Scented bleaches can mask one's natural ability to detect these harmful fumes.

The Chemical Details

Ammonia (NH_3) and hypochlorite ion (OCl^-) combine to produce three different chloramine species — that is, compounds that are derivatives of ammonia in which one or more of the hydrogen atoms has been replaced by a chlorine atom. In order of increasing degree of chlorine substitution, these chloramines are named monochloramine (NH_2Cl), dichloramine ($NHCl_2$), and nitrogen trichloride (NCl_3):

$$OCl^- + NH_3 \rightarrow OH^- + NH_2Cl$$
$$OCl^- + NH_2Cl \rightarrow OH^- + NHCl_2$$
$$OCl^- + NHCl_2 \rightarrow OH^- + NCl_3.$$

The pungent and irritating odor of chloramines is often mistaken for the "chlorine odor" of swimming pools. Chloramines form from the combination of sodium hypochlorite (added to sterilize the water) and nitrogen-containing compounds that are human waste by-products.

In the presence of excess ammonia, hypochlorite ion and ammonia can combine to form hydrazine (N_2H_4), another toxic and potentially explosive substance:

$$OCl^- + NH_3 \rightarrow NH_2Cl + OH^-$$
$$NH_2Cl + NH_3 + OH^- \rightarrow N_2H_4 + Cl^- + H_2O.$$

Industrial preparation of hydrazine is based on this reaction of ammonia with an alkaline solution of sodium hypochlorite, known as the *Raschig process* introduced in 1907.

KEY TERMS: chloramines hydrazine

Related Web Sites

► "Pool Water Chemistry: Technical Details." Virtual Pool & Spa Store, Long Island Hot Tubs & Paramount Pools, http://paramountpools.com/page371.htm

13.10 | What Is "Alpha Hydroxy Acid" in Antiaging Creams?

Many major cosmetic companies have marketed "antiaging" creams and skin-renewal products containing alpha hydroxy acids as the miracle ingredients. These particular components are easily identified in a list of a lotion's ingredients, provided you know some simple chemistry nomenclature.

The Chemical Basics

Many antiwrinkle creams and lotions containing alpha hydroxy acids claim to reverse the effect of sun damage and aging on skin. Most of these products improve the skins appearance and texture by accelerating the natural process by which the skin replaces its aging outer layer (the epidermis) with new cells. By causing the skin to peel, alpha hydroxy acids temporarily enhance the skin's appearance by revealing "healthier-looking" skin in the lower layers. The chemical identity of the antiaging ingredient, its concentration, its pH or acidity, and the presence of other ingredients in the skin product determine the effectiveness of the acid as an exfoliator.[1]

Alpha hydroxy acids (AHAs) are water-soluble substances and thereby penetrate the outermost epidermal skin layers.[2] In contrast, beta hydroxy acids (BHAs) are lipid (fat) soluble and are capable of penetrating to the underlying layers of skin (the dermis) located 1–5 mm below the surface of the skin.[2] Most AHAs are derived from plant materials and marine sources. Commonly used AHAs include malic acid (found in apples), ascorbic acid (a common ingredient in numerous fruits), glycolic acid (a constituent of sugar cane), lactic acid (a component of milk), citric acid (naturally abundant in citrus fruits), and tartatic acid (found in red wine). A common BHA is salicylic acid (an ingredient in aspirin).

The Chemical Details

An alpha hydroxy acid is an organic carboxylic acid in which an additional hydroxyl functional group (–OH) is present at the alpha position, i.e., on the carbon adjacent to the carboxyl functionality, –COOH. Figure 13.10.1 presents the struc-

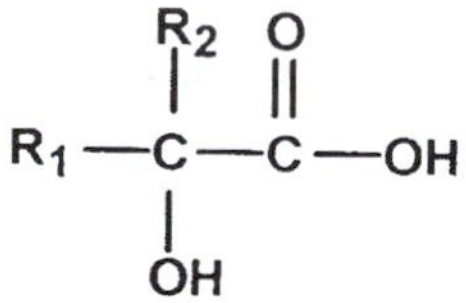

$$R_1 - \overset{\overset{\displaystyle R_2}{|}}{\underset{\underset{\displaystyle OH}{|}}{C}} - \overset{\overset{\displaystyle O}{||}}{C} - OH$$

Figure 13.10.1 ▶ The structure of an alpha hydroxy acid, an organic carboxylic acid in which an additional hydroxyl functional group (–OH) is present at the alpha position, i.e., on the carbon adjacent to the carboxyl functionality, –COOH.

ture of a generic AHA. A beta hydroxy acid is also a substituted carboxylic acid in which the hydroxyl substituent is attached to the beta position, i.e., at a carbon atom two positions away from the carboxylic acid functionality, given by the structure in Fig. 13.10.2. Some common alpha hydroxy acids are glycolic acid

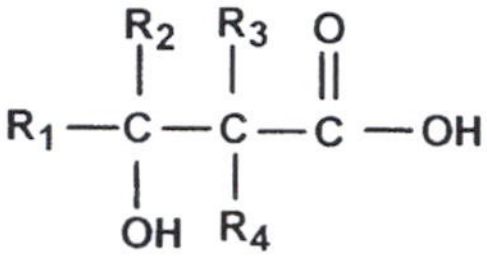

$$R_1 - \overset{\overset{\displaystyle R_2}{|}}{\underset{\underset{\displaystyle OH}{|}}{C}} - \overset{\overset{\displaystyle R_3}{|}}{\underset{\underset{\displaystyle R_4}{|}}{C}} - \overset{\overset{\displaystyle O}{||}}{C} - OH$$

Figure 13.10.2 ▶ The structure of a beta hydroxy acid, a substituted carboxylic acid in which the hydroxyl substituent is attached to the beta position, i.e., at a carbon atom two positions away from the carboxylic acid functionality.

or α-hydroxyethanoic acid (Fig. 13.10.3), lactic acid (Fig. 13.10.4), malic acid

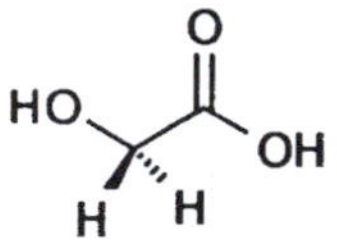

Figure 13.10.3 ▶ The molecular structure of glycolic acid or α-hydroxyethanoic acid.

(Fig. 13.10.5), citric acid (Fig. 13.10.6), mandelic acid (Fig. 13.10.7), ascorbic acid (Fig. 13.10.8), and tartaric acid (Fig. 13.10.9). Salicylic acid (Fig. 13.10.10) is the most common beta hydroxy acid, although citric acid could also be classified as a BHA.

KEY TERMS:	carboxylic acid	carboxyl functionality
	beta hydroxy acid	alpha hydroxy acid

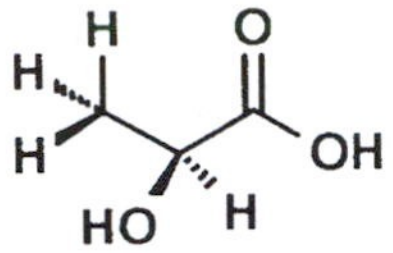

Figure 13.10.4 ▶ The molecular structure of lactic acid.

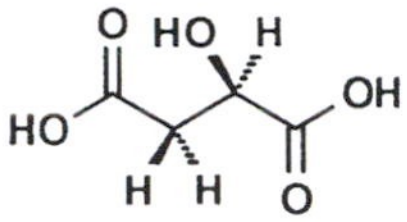

Figure 13.10.5 ▶ The molecular structure of malic acid.

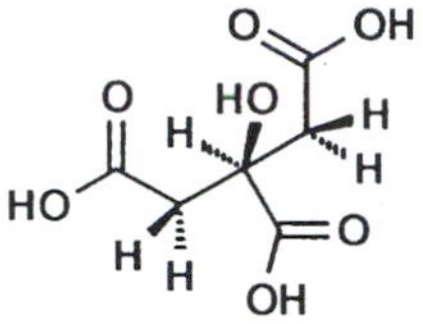

Figure 13.10.6 ▶ The molecular structure of citric acid.

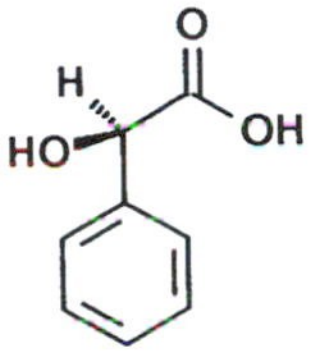

Figure 13.10.7 ▶ The molecular structure of mandelic acid.

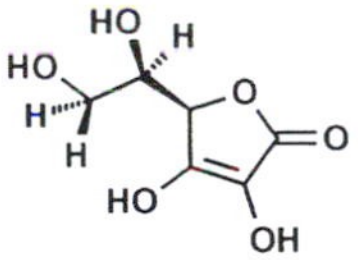

Figure 13.10.8 ▶ The molecular structure of ascorbic acid.

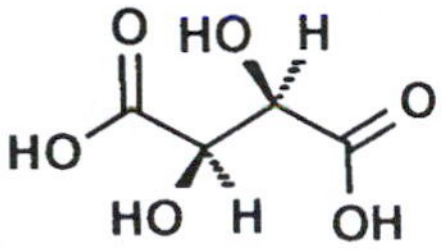

Figure 13.10.9 ▶ The molecular structure of tartartic acid.

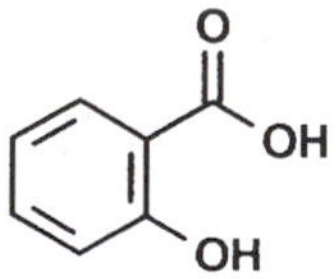

Figure 13.10.10 ▶ The molecular structure of salicylic acid.

References

[1] "Current Controls." Australia's National Industrial Chemicals Notification and Assessment Scheme, http://www.nicnas.gov.au/publications/CAR/PEC/PEC12/PEC12c.pdf

[2] "Hydroxy Acids." YourSkinDoctor.com, Inc., http://www.yourskindoctor.com/hydroxy_acids.html

Related Web Sites

▶ "Alpha Hydroxy Acids in Cosmetics." U.S. Food and Drug Administration, *FDA Backgrounder*, July 3, 1997, http://vm.cfsan.fda.gov/~dms/cos-aha.html

▶ "Alpha Hydroxy Acids for Skin Care: Smooth Sailing or Rough Seas?" Paula Kurtzweil, U.S. Food and Drug Administration, FDA Consumer, March–April 1998, http://vm.cfsan.fda.gov/~dms/fdacaha.html

13.11 What Is an "Alcohol-Free" Cosmetic?

Before you reach for that "alcohol-free" product, stop and consider the features that you expect and desire in your cosmetics. The class of substances known to chemists as alcohols is commonly confused with the consumer interpretation of "alcohol." Knowing the chemist's terminology will help you select the safest and most beneficial formulation.

The Chemical Basics

The U.S. Food and Drug Administration reports that consumers often inquire about the meaning of the phrase "alcohol-free" when describing a cosmetic product. The term "alcohol-free" specifically applies to the absence of the substance known as *ethanol* or *ethyl alcohol* or *grain alcohol*.[1] An alcohol-free product is often desired by users wishing to avoid drying effects often attributed to many alcohol-containing products. Nevertheless, many desirable features of cosmetics — moisturizing, smoothness, emolliency, ease of application, fast drying, etc. — arise from the presence of substances that belong to the broader chemical family of substances known simply as alcohols.

The Chemical Details

An alcohol is a member of a class of chemical substances that contain the functional group known as the *hydroxyl* group. This functionality is designated –O–H, i.e., a hydrogen atom bonded to an oxygen atom that is subsequently bonded to another atom, in particular, carbon. Cosmetic products, whether labeled "alcohol-free" or not, may contain members of the alcohol family.[2,3] Alcohols confer many highly desirable characteristics to personal care products. For example, many alcohols are key ingredients to act as moisturizers, including cetyl alcohol or hexadecanol (Fig. 13.11.1) in facial make-up, hair products, and deodorants;

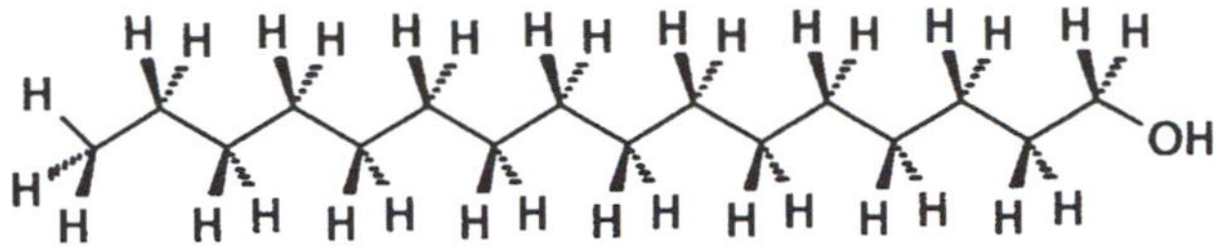

Figure 13.11.1 ▶ The molecular structure of cetyl alcohol or hexadecanol.

2-octyl-1-dodecanol (Fig. 13.11.2) in skin conditioners; panthenol (Fig. 13.11.3)

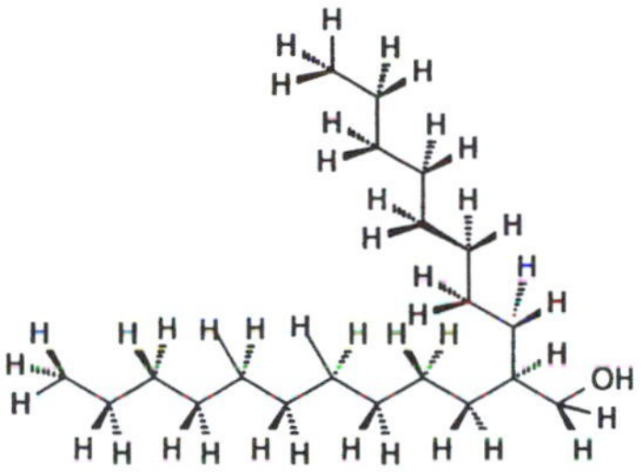

Figure 13.11.2 ▶ The molecular structure of 2-octyl-1-dodecanol.

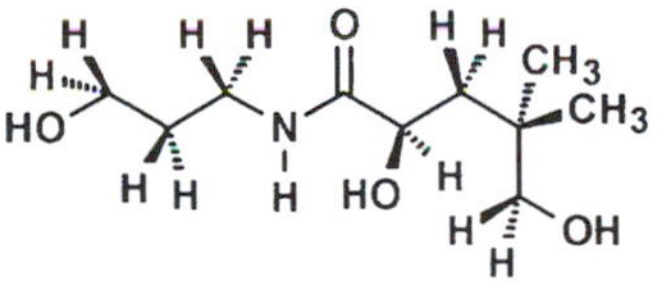

Figure 13.11.3 ▶ The molecular structure of panthenol.

in hair conditioners contain stearyl alcohol or octadecanol (Fig. 13.11.4), and lanolin alcohols (a mixture of 33 high-molecular-weight alcohols) are found in skin and hair conditioners. The alcohol α-tocopherol (Vitamin E) (Fig. 13.11.5) is used as an antioxidant and preservative to prevent or reduce product deterioration. Phenoxyethanol (Fig. 13.11.6) is used in synthetic rose oils and soaps as the rose fragrance component. Many alcohols are included for the solvency

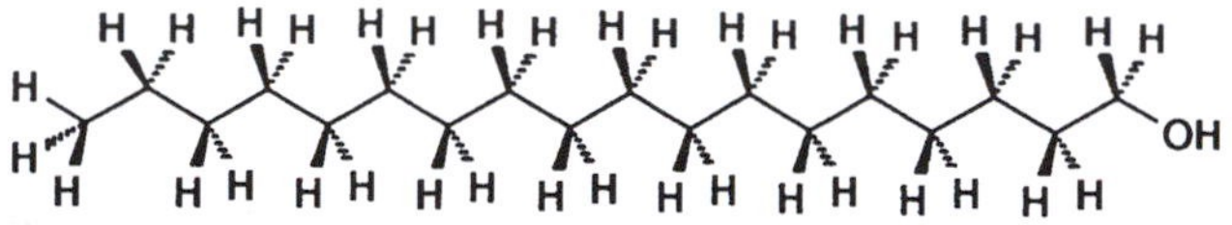

Figure 13.11.4 ▶   The molecular structure of stearyl alcohol or octadecanol.

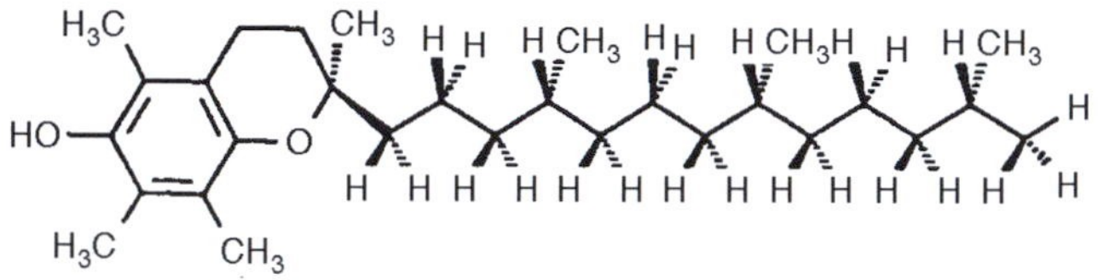

Figure 13.11.5 ▶   The alcohol α-tocopherol known as vitamin E.

properties, that is, their ability to dilute the formulation to the desired consistency. In addition to ethanol (Fig. 13.11.7), the alcohols employed for this purpose are substances with more than one hydroxyl group, including glycerin or 1,2,3-propanetriol (Fig. 13.11.8), propylene glycol or 1,2-propanediol (Fig. 13.11.9), and butylene glycol or 1,4-butanediol (Fig. 13.11.10). Lauric acid diethanolamine or lauramide DEA (Fig. 13.11.11) is often included for its emulsifying properties and foam action. Many permanent hair dye formulations contain derivatives of aminophenol or hydroxyaniline (Fig. 13.11.12). Triethanolamine (Fig. 13.11.13) is an alcohol included as a pH adjuster in transparent soap. The functions and benefits of alcohols in cosmetic formulations are endless! Users of cosmetic products should probably be quite relieved to learn that their products are not completely hydroxyl free.

KEY TERMS: alcohol   hydroxyl group

References

[1] "Alcohol Free." U.S. Food and Drug Administration, Center for Food Safety and Applied Nutrition, Office of Cosmetics Fact Sheet, February 23, 1995,
http://vm.cfsan.fda.gov/~dms/cos-227.html

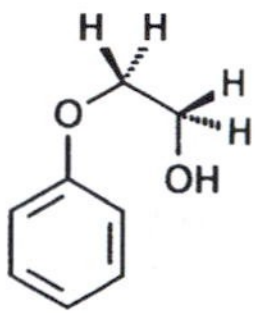

Figure 13.11.6 ▶   The molecular structure of phenoxyethanol.

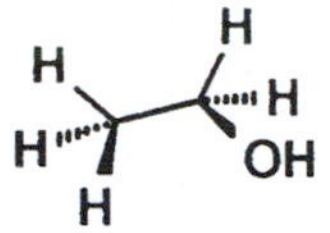

Figure 13.11.7 ▶ The molecular structure of ethanol.

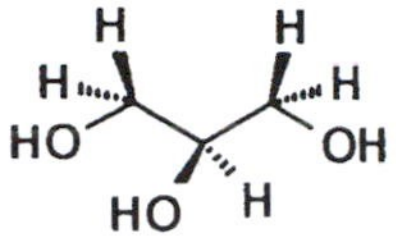

Figure 13.11.8 ▶ The molecular structure of glycerin or 1,2,3-propanetriol.

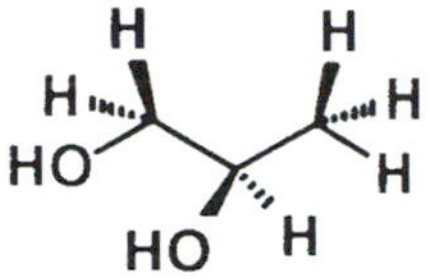

Figure 13.11.9 ▶ The molecular structure of propylene glycol or 1,2-propanediol.

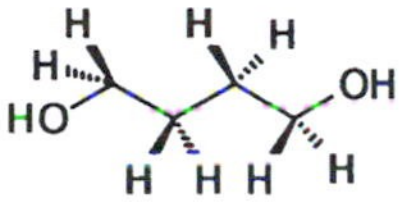

Figure 13.11.10 ▶ The molecular structure of butylene glycol or 1,4-butanediol.

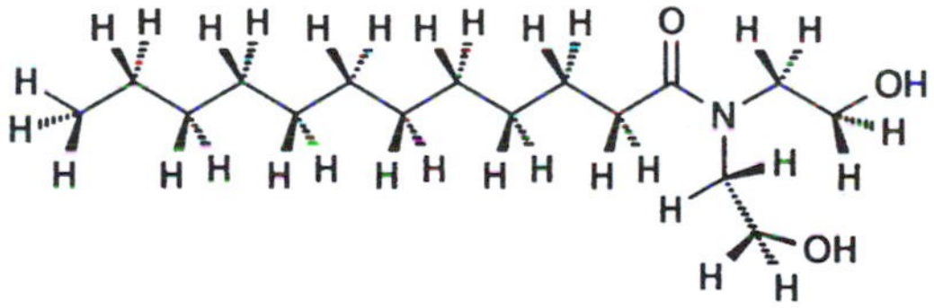

Figure 13.11.11 ▶ The molecular structure of lauric acid diethanolamine or lauramide DEA.

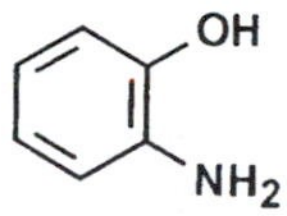

Figure 13.11.12 ▶ The molecular structure of aminophenol or hydroxyaniline.

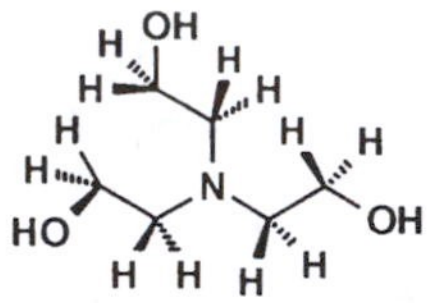

Figure 13.11.13 ▶   The molecular structure of triethanolamine.

[2] "Hydroalcoholic Skin Care Emulsions." Uniqema,
http://www.surfactants.net/formulary/uniqema/pcm14.html

[3] "Chemical Ingredients Found in Cosmetics." U.S. Food and Drug Administration, *FDA Consumer*, May 1994. http://vm.cfsan.fda.gov/~dms/cos-chem.html

13.12 What Is the "Cool Sensation" in Toothpaste and Breath Fresheners?

"Invigorating mint taste for a clean, healthy mouth and fresh breath that lasts." "Clean, cool, refreshing taste." These claims accompany many advertisements for toothpastes and breath fresheners. What is the source of the refreshing flavor and cooling effect that consumers readily associate with these products?

The Chemical Basics

We often experience a cooling, refreshing sensation when we use various tooth-pastes and mouthwashes. Manufacturers of such personal hygiene products include one or more key ingredients to provide both a pleasant taste and a bracing feeling of coolness. These components include peppermint oils, spearmint oils, and menthol. These substances belong to a class of materials known as *essential oils* — the highly concentrated, volatile, aromatic essences of specific plants. To acquire these key substances, the flowers, stems, bark, and leaves of plants must be harvested at exact growth stages, under certain weather conditions, and even at specific times of the day.[1] Specific constituents in these essential oils stimulate the nerves that sense cold and depress those nerves that detect pain. This sensory process is known as *chemoreception* and is initiated when certain chemical stimuli come in contact with chemoreceptors, those specialized cells in the body that convert the immediate effects of such substances directly or indirectly into nerve impulses. The body subsequently responds, producing its own "warming effect" as blood flows into the area of application. This physical sensation impresses the senses as a "medicinal" effect and is partially responsible for peppermint's long history of use as medicine.

The Chemical Details

Peppermint oil is the volatile oil extracted from the fresh leaves of the flowering plant of *Mentha piperita* via steam distillation. Peppermint oil contains not less than 44% menthol. American peppermint oil contains from 50 to 78% of free *l*-menthol (Fig. 13.12.1) and from 5 to 20% combined in various es-

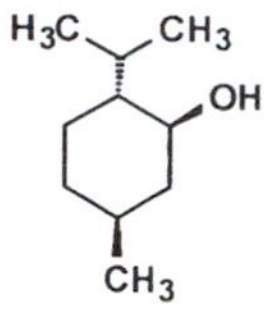

Figure 13.12.1 ▶ The molecular structure of *l*-menthol.

ters such as menthyl acetate (see Fig. 13.12.2). The essential oil also contains

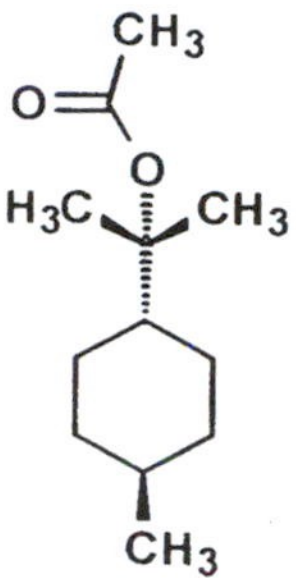

Figure 13.12.2 ▶ The molecular structure of menthyl acetate.

d-menthone, *l*-menthone (Fig. 13.12.3), cineole (Fig. 13.12.4), *d*-isomenthone, *d*-

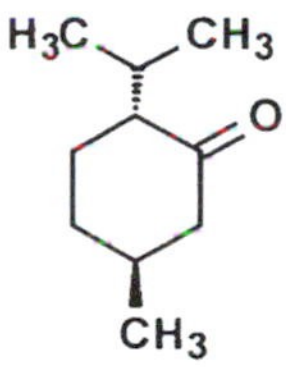

Figure 13.12.3 ▶ The molecular structure of *l*-menthone.

neomenthone, and the monoterpene derivative menthofuran (Fig. 13.12.5).[2] The alcohol *l*-menthol and the ester menthyl acetate are responsible for the pungent and refreshing odor associated with peppermint oil. The specific constituents are found to depend on the age of the leaves, with menthol and menthyl acetate preferably found in older leaves and preferentially formed during long daily sunlight periods. The ketones menthone (Fig. 13.12.3) and pulegone (Fig. 13.12.6) (and

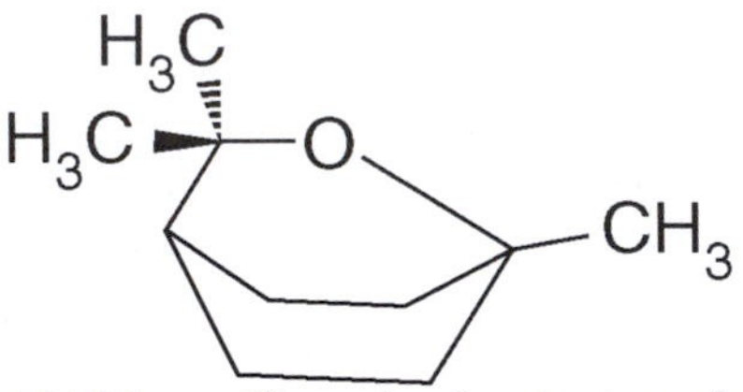

Figure 13.12.4 ▶ The molecular structure of cineole.

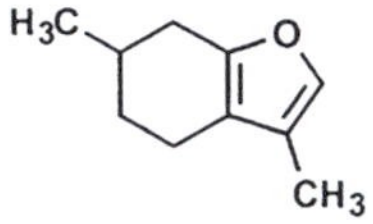

Figure 13.12.5 ▶ The molecular structure of the monoterpene derivative menthofuran.

the monoterpene derivative menthofuran (Fig. 13.12.5)) have a less delightful fragrance and appear to a greater extent in young leaves. The formation of these latter ingredients occurs predominantly during short days. The cooling sensation is associated with the *l*-menthol optical isomer, i.e., the levorotatory form that rotates the plane of polarized light to the left. Recall that *optical isomerism* results when two structures are identical in composition and bonding but are nonsuperimposable mirror images of each other. As a consequence of their three-dimensional structure, optical isomers rotate the plane of polarization of a beam of polarized light that is directed through them in different directions. The levorotatory or *l* form rotates the plane of polarized light to the left; the dextrorotatory or *d* form rotates the plane of polarized light to the right.

The main constituents of spearmint oil are *l*-carvone (Fig. 13.12.7) and *l*-limonene (Fig. 13.12.8). Oil of spearmint contains from 45 to 60% *l*-carvone, 6 to 20% of alcohols, and 4 to 20% of esters and terpenes, mainly *l*-limonene and cineole (see Fig. 13.12.4).[2] The optically isomeric form of carvone, *d*-carvone, is found in oil of caraway and oil of dill. Carvone appears to co-occur with limonene when present in a plant.

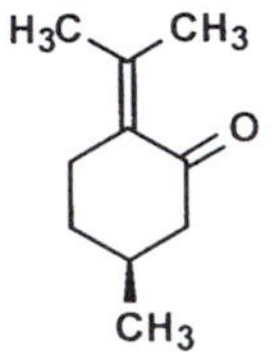

Figure 13.12.6 ▶ The molecular structure of pulegone.

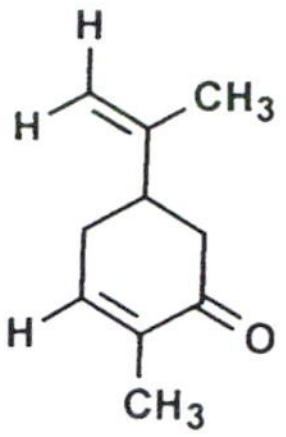

Figure 13.12.7 ▶ The molecular structure of *l*-carvone.

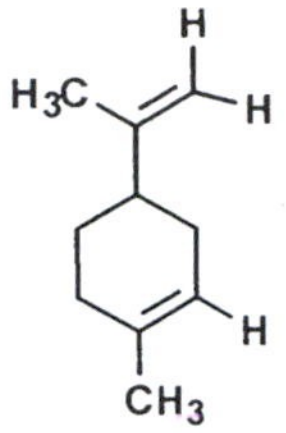

Figure 13.12.8 ▶ The molecular structure of *l*-limonene.

KEY TERMS: optical isomers

References

[1] "About Aromatherapy: Aromatherapy Guide." Ashbury's Aromatherapy, http://www.ashburys.com/exp_aroma_guide.html

[2] "Structure/Activity Comparison in the Ability of Some Terpenoid Food Flavours to Cause Peroxisome Proliferation." "Literature References," Georges-Louis Friedli, September 1992, University of Surrey, School of Biological Sciences, http://friedli.com/research/MSc/literature.html

Related Web Sites

▶ "Essential Oils Industry." Alberta: Agriculture, Food, and Rural Development, http://www.agric.gov.ab.ca/agdex/100/8883001.html

▶ "Identification of the Volatile and Semi-Volatile Organics in Chewing Gums by Direct Thermal Desorption." John J. Manura, Scientific Instrument Services, Inc., Short Path Thermal Desorption — Application Note No. 12 — October 1992, http://www.sisweb.com/referenc/applnote/ap12-a.htm

13.13 When We Buy Fresh Fish, Why Does Its Smell Indicate Its Freshness?

When someone buys fresh fish, they often smell the fish to test for freshness — the fresher the fish, the less "fishy" the smell. What is the source of the fishy odor?

The Chemical Basics

The unpleasant smell of rotten fish is a consequence of natural metabolic processes that occur as the fish decay. Specifically, the bacterial enzymes that decompose the flesh of fish degrade one of the natural constituents of fish muscle, trimethylamine oxide (TMAO), to odiferous products. Marine fish, shellfish, and some freshwater fish contain TMAO, presumably a component of their osmoregulatory system that is important for maintaining water balance in saltwater.[1] High concentrations of TMAO have also been correlated with deep-water fish, potentially aiding the stabilization of the proteins in the fish against the high pressures experienced at great depths.[2] The decomposition products of TMAO include trimethylamine and dimethylamine. A "fishy" smell is nothing more than the odor of a mixture of amines. The amount of trimethylamine formation is often used as an index of fish quality.

To remove a fishy smell, wash with a solution of baking soda, lemon juice, or mild vinegar. Why? These substances act as acids to neutralize the amines, which are alkaline (basic) substances.

The Chemical Details

Metabolism of trimethylamine oxide in fish muscle involves an enzyme-catalyzed oxidation–reduction reaction. The enzyme responsible for the conversion of trimethylamine oxide to trimethylamine is known as trimethylamine-N-oxide reductase. This enzyme acts on nicotinamide adenine dinucleotide (NADH) and TMAO to produce NAD^+, trimethylamine and water (Fig. 13.13.1). TMAO acts as the oxidizing agent and is reduced, while NADH undergoes oxidation as the reducing agent.

An alternate enzyme, trimethylamine-oxide aldolase, converts TMAO to dimehylamine and formaldehyde, as written in Fig. 13.13.2.

> **KEY TERMS:**   amine   acid–base neutralization

References

[1] "Rotten Fish." New York Times on the Web, *Science Health*, May 23, 2000, http://www.nytimes.com/library/national/science/052300sci-qa.html

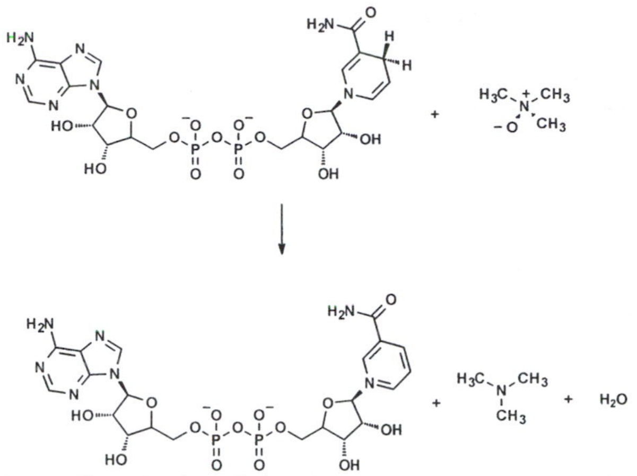

Figure 13.13.1 ► The reduction of trimethylamine oxide by nicotinamide adenine dinucleotide NADH.

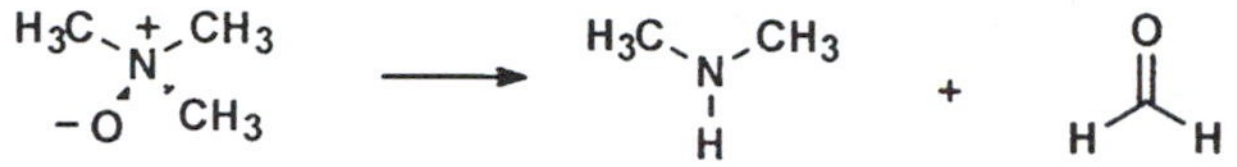

Figure 13.13.2 ► Alternative reduction pathway of trimethylamine oxide to dimethylamine and formaldehyde.

[2] "Osmolytes in the Deep." Professor Paul H. Yancey, Whitman University, http://people.whitman.edu/~yancey/news.html

[3] Ligand Chemical Database, GenomeNet WWW Server, Institute for Chemical Research, Kyoto University, http://www.genome.ad.jp/dbget-bin/www_bget?ligand+1.6.6.9

[4] Ligand Chemical Database, GenomeNet WWW Server, Institute for Chemical Research, Kyoto University, http://www.genome.ad.jp/dbget-bin/www_bget?ligand+4.1.2.32

Related Web Sites

► "Blowing the Theory of How We Smell." David Bradley, Elemental Discoveries, Summer 1997, ScienceBase.com, http://www.sciencebase.com/elemsmell.html

► "Methane Metabolism." Ligand Chemical Database, GenomeNet WWW Server, Institute for Chemical Research, Kyoto University, http://www.genome.ad.jp/dbget-bin/show_pathway?MAP00680+C01104

13.14 Are Flamingos Naturally Pink?

The flamingo is often associated with the ancient Phoenix, an immortal bird that was consumed by flames only to rise from the ashes. The word "flamingo" is also associated with fire or flames, derived from the Portuguese *flamengo* or Spanish *flamenco* and with the Latin root of *flamma*, flame. How do flamingos come by the reddish-pinkish plumage for which they are named?

The Chemical Basics

Bird feathers exhibit a rich array of coloring. Some feather colors arise from pigments in the feather, such as the red color of a cardinal. Other coloring arises from variations in the feather structure. For example, microscopic structures on the feather surface of the blue Scrub Jay act as tiny prisms to reflect light to create the blue appearance. For flamingos, the bird's diet is responsible for the pigmentation found in the feathers with pink coloration. Specifically, cartenoids are a family of colored substances that include β-carotene in carrots and beets, lycopene in tomatos and pink grapefruit, and the colorants in salmon flesh or the shells of shrimp and lobster. While young flamingos have white coloring, the diet of adult flamingos includes carotenoid-containing small crustaceans, insects, and red algae. The colored substance is ingested and modified to give rise to the pink shade in flamingo feathers. A continual diet of carotenoid-containing food is necessary to maintain the pink coloring, for captive flamingos often lose their bright colors without special dietary supplements. During the feeding of their young chicks, both the male and female parent flamingos deplete their stores of the pink-colored carotenoid canthaxanthin to such an extent that, if molting occurs after nesting, the new feathers come in white instead of pink.[1]

The Chemical Details

One member of the carotenoid family, astaxanthin (Fig. 13.14.1), is responsi-

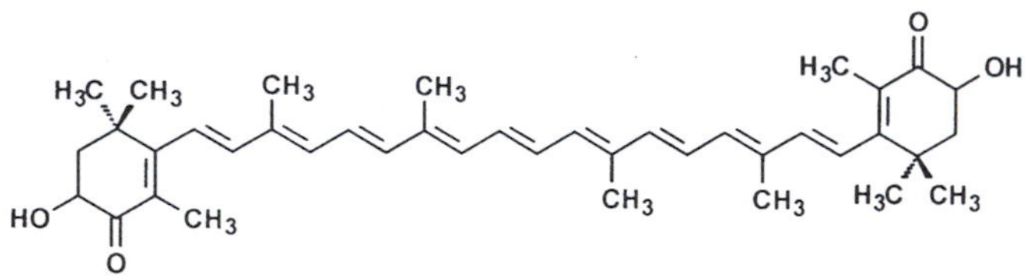

Figure 13.14.1 ▶ The molecular structure of astaxanthin.

ble for the pink color of salmon, shellfish, shrimp, and lobster. After ingestion, this compound is modified to remove the two hydroxyl groups, producing canthaxanthin (Fig. 13.14.2), the actual substance that gives the flamingo its pink color.[1] Canthaxanthin in cyclohexane solvent principally absorbs light of wave-

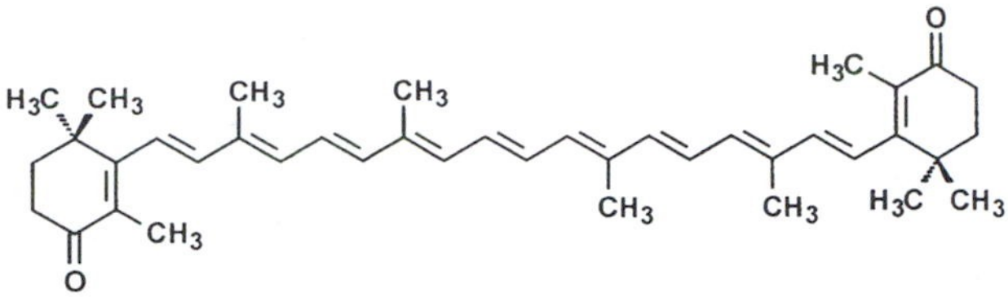

Figure 13.14.2 ▶ The molecular structure of canthaxanthin.

length around 468–472 nm[2] in the green region of the visible spectrum, and thus reflects the red color that we associate with this pigment.

References

[1] "Behavior — Ecological and Evolutionary Implications." Biology 213, Introduction to Ecology and Evolutionary Biology, Rice University,
http://www.dacnet.rice.edu/courses/bios213/labs/flamingo/

[2] "Food Colorants/Carotenoids: Structure and Solubility." Food, Nutrition, and Health 410, University of British Columbia, http://www.agsci.ubc.ca/courses/fnh/410/colour/3_30.htm

Related Web Sites

▶ "How Do Feathers Get Their Colors?" The Bird Site, Natural History Museum of Los Angeles County, http://www.lam.mus.ca.us/birds/guide/pg012.html

▶ "Flamingos — Physical Characteristics." Sea World Adventure Parks, Busch Gardens, http://www.shamutv.com/Flamingos/fphysical.html

▶ "Carotenoids." Willi Stahl, http://www.uni-duesseldorf.de/WWW/MedFak/PhysiolChem/_fields/carot.html

13.15 | What Causes Ice Cream to Develop a Gritty Texture during Long Periods of Storage?

The texture of ice cream is an important quality factor to consumers. What is the chemical basis for the gritty or sandy-like texture that detracts from ice cream's pleasant taste?

The Chemical Basics

Carbohydrates were first analyzed in the early nineteenth century and described as compounds that were literally "hydrates of carbon" because they had the general formula $C_x(H_2O)_x$. In recent years, carbohydrate molecules are generally classified on the basis of their structures, not their formulas. Among the compounds that belong to this family are cellulose, starch, glycogen, and most sugars.

Carbohydrates may be classified as monosaccharides (simple carbohydrates that cannot be hydrolyzed, i.e., decomposed in water), disaccharides (carbohydrates that hydrolyze to yield two monosaccharides), or even more complex polysaccharides containing as many as thousands of monosaccharides linked together. The sugar found in milk, lactose, is a disaccharide consisting of a galactose molecule and a glucose molecule. In solution, each of these monosaccharide units is present as a ring structure. Two structures, each of different spatial orientation (i.e., different three-dimensional arrangements of the atoms), are observed, and each structure exhibits different physical properties. One of the structures has a sweeter taste, but the other structure has a lower solubility. When ice cream is stored over a longer time, the less soluble structure crystallizes from the mixture, leading to a gritty texture.

The Chemical Details

The ring structures of monosaccharides can lead to two orientations of one of the –OH groups relative to the ring. These structures are known as anomers and designated α and β. The α designation is used when the –OH group on carbon C1 is below the ring and the –CH$_2$OH substituent on carbon C5 is above the ring, and the β distinction is used when the –OH group and the –CH$_2$OH substituent are on the same side of the ring. Carbon C1 is known as the *anomeric* carbon. The α and β forms are in equilibrium with one another and interconvert. The α and β forms of D-glucose are illustrated in Figs. 13.15.1 and 13.15.2, respectively.

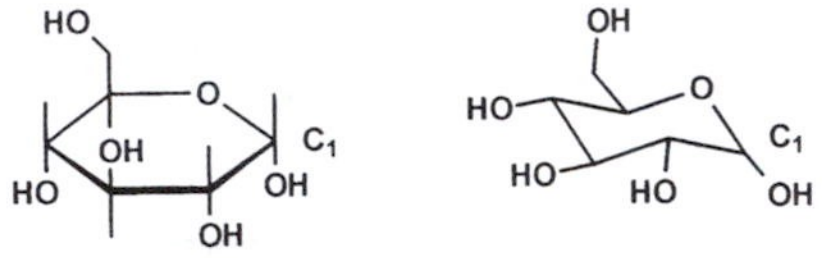

Figure 13.15.1 ▶ The molecular structure of α-D-glucose.

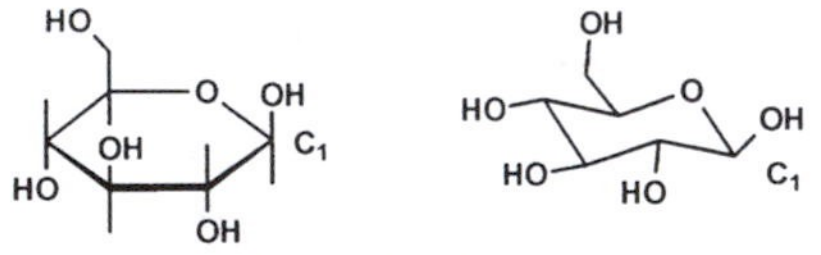

Figure 13.15.2 ▶ The molecular structure of β-D-glucose.

The disaccharide lactose is formed by the reaction of the –OH group on the anomeric carbon of β-D-galactose (i.e., C1 on β-D-galactose) with an –OH group on the C4 carbon of either α- or β-D-glucose. Thus two forms of lactose are possible and identified as α-lactose and β-lactose (Figs. 13.15.3 and 13.15.4, respectively), depending on the orientation of the hydroxyl group on the anomeric carbon in the glucose unit.

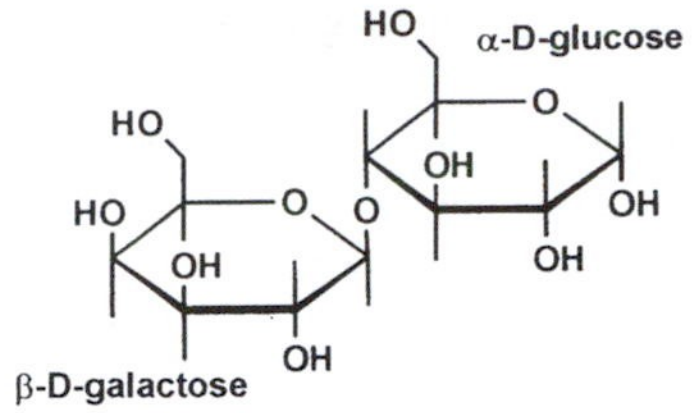

Figure 13.15.3 ► The molecular structure of α-lactose.

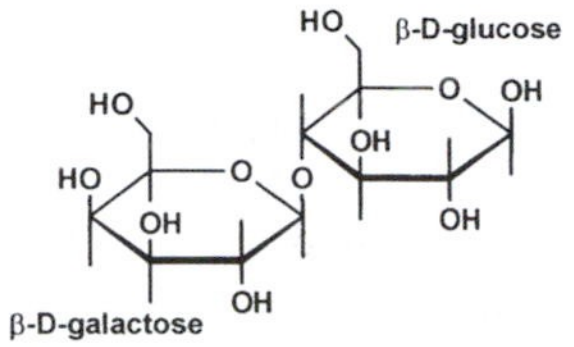

Figure 13.15.4 ► The molecular structure of β-lactose.

The linkage between the monosaccharide units is an ether linkage –O– known as a *glycosidic linkage* and is further named according to the number of the carbon at which the linkage begins and the carbon on the second monosaccharide at which the linkage ends. In lactose the glycosidic linkage is designated as a 1,4-glycosidic bond.

KEY TERMS: ethers   carbohydrates

Related Web Sites

► "Carbohydrates." Illinois State University, http://xenon.che.ilstu.edu/genchemhelphomepage/topicreview/bp/1biochem/carbo5.html

► "Sandiness in Ice Cream." Dairy Research and Information Center, University of California — Davis, http://drinc.ucdavis.edu/html/icream/icream-3.shtml

13.16 Why Is Some Olive Oil Designated as "Extra Virgin"?

For culinary experts and consumers alike, the words "extra virgin" have a ring of exclusiveness and quality when it comes to olive oil. What does the term actually signify?

TABLE 13.1 ▶ Classification of Virgin Olive Oils by Composition[1,3]

Extra virgin	Acidity less than 1%
virgin	Acidity less than 2%
Ordinary virgin	Acidity less than 3.3%
Lampante virgin	More than 3.3%, which is not suitable for consumption without refining

The Chemical Basics

Edible oils from plants are sold and consumed as either pure oils from a single plant or as blended oils. Olive, sunflower, and corn oils fall into the first category; oils designated as "edible," "cooking," "frying," "table," or "salad" oil are examples of the second classification. Olive oil is considered by epicureans and gourmet cooks to be a superior vegetable oil for its desirable taste and aroma. Virgin olive oils must be physically extracted with special mechanical and thermal conditions that do not lead to deterioration. Virgin olive oils are further classified according to four grades.[1] Acidity is one of the factors defining the grades (see Table 13.1).

The highest grade, "extra virgin" olive oil, must have an acidity of less than 1% by weight. In addition, to receive the designation of "extra virgin" an oil must be further judged by a tasting panel of the International Olive Oil Council to have an intense fruit flavor and distinctive aroma and taste. Lower grade oils might undergo extra processing to neutralize acid or to eliminate an unappealing odor.

By the nature of the process by which olive oil is extracted from the olive, the oil is susceptible to contamination. The high price associated with olive oil of the highest purity—"extra virgin olive oil"—also leads to falsification by unscrupulous vendors who blend with less costly oils such as corn, peanut, and soybean oil. Various analytical techniques have been devised to authenticate the purity of olive oil by detecting certain oil components.

The oil present in olives is extracted in a complex separation process.[2] After removing twigs and stems, olives are first washed and then ground and squeezed into paste. The olive paste is then mixed at 28°C for 20–40 min to allow small oil droplets to coalesce into larger droplets. A device known as a centrifugal decanter is then used to separate the water and oil in the fruit from the fruit flesh and pits. This piece of equipment consists of a spinning horizontal drum; centrifugal force directs the heavier flesh and pits to the outside and the water and oil are tapped off from the center. To separate the oil from the water, either a centrifuge or a decanter is employed. With a centrifuge, the spinning liquid separates the heavier water from the oil. Gravity separates the oil and water in a decanter. Alternatively, a simple separation process involves mixing the stirred olive paste with large quantities of warm water. The oil floats to the top surface and is removed with a skimmer device. In either situation, additives such as hexane or other solvents, salt (sodium chloride), or alkaline substances may be included to extract additional oil from the fruit.

The Chemical Details

How do olive oils vary in acid content? Oils are a complex mixture of compounds known as triglycerides. Triglycerides are esters formed from the alcohol glycerol (1,2,3-propanetriol, $C_3H_8O_3$, Fig. 13.16.1) and three carboxylic acids

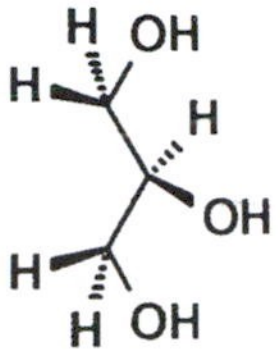

Figure 13.16.1 ▶ The molecular structure of glycerol (glycerin).

known as *fatty acids*. Fatty acids generally contain a long hydrocarbon chain (typically with an even number of carbon atoms ranging between 12 and 24) and a carboxylic acid group, –COOH. A large number of fatty acids exist as a consequence of variations in the chain length and the degree of saturation (i.e., number and placement of carbon–carbon double bonds, see below). Figure 13.16.2 illus-

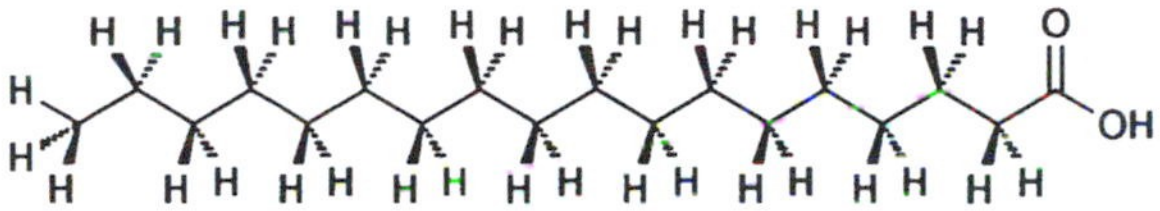

Figure 13.16.2 ▶ The molecular structure of stearic acid.

trates a typical fatty acid. The fatty acids of a triglyceride may have the same or different identities. A simple triglyceride contains three identical fatty acids; a mixed triglyceride contains at least two different fatty acids. Figure 13.16.3 illustrates the chemical structure of a simple triglyceride composed of three identical fatty acids, oleic acid. Figure 13.16.4 illustrates a mixed triglyceride, palmitodiolein; as the name implies, this triglyceride is composed of one molecule of palmitic acid and two molecules of oleic acid. In humans triglycerides are an important source of fuel generally obtained from food. As triglycerides are found in the bloodstream and stored in fat (adipose) tissue, elevated levels of triglycerides, however, are associated with an increased risk of heart disease. Esters undergo hydrolysis, particularly in the presence of certain enzymes or with the use of alkaline substances and heat, to decompose to yield the constituent alcohol and carboxylic acid. Triglycerides hydrolyze to yield glycerol and the component fatty acids (Fig. 13.16.5). The hydrolysis of a triglyceride is also known as

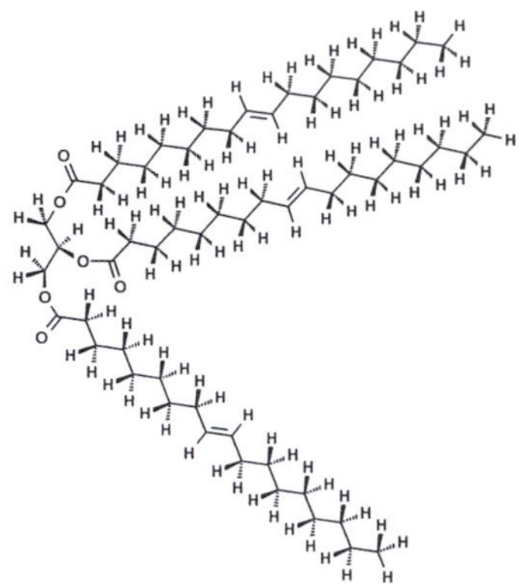

Figure 13.16.3 ▶ The molecular structure of a triglyceride, a fatty acid based on the triester of oleic acid.

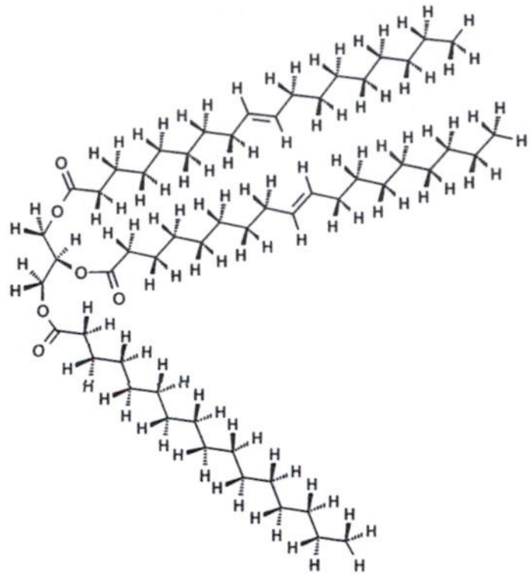

Figure 13.16.4 ▶ The molecular structure of the mixed fatty acid palmitodiolein, which is triglyercide of two esters of oleic acid and one of palmitic acid.

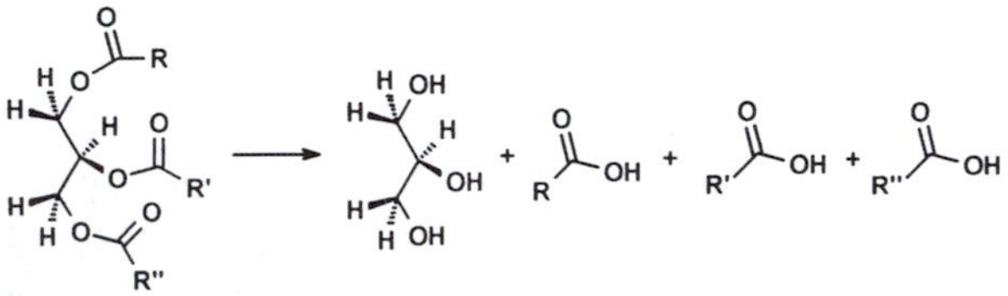

Figure 13.16.5 ▶ The hydrolysis or saponification of a triglyceride to yield the constituent acids and glycerol.

TABLE 13.2 ▶ Common Fatty Acids in Oils from Plants

Common name / Systematic Name	No. of Carbons	Structural formula
<Saturated>		
Lauric	12	$CH_3(CH_2)_{10}COOH$
n-Dodecaneic		
Myristic	14	$CH_3(CH_2)_{12}COOH$
n-Tetradecanoic		
Palmitic	16	$CH_3(CH_2)_{14}COOH$
n-Hexadecanoic		
Stearic	18	$CH_3(CH_2)_{16}COOH$
n-Octadecanoic		
<Unsaturated>		
Oleic	18	$CH_3(CH_2)_7CH{=}CH(CH_2)_7$
cis-9-Octadecenoic		$COOH$ (*cis*)
Linoleic	18	$CH_3(CH_2)_4CH{=}CHCH_2CH{=}$
cis,cis-9,12-Octadecadienoic		$CH(CH_2)_7COOH$ (*cis, cis*)
Linolenic	18	$CH_3CH_2CH{=}CHCH_2CH{=}$
cis,cis,cis-9,12,15-		$CHCH_2CH{=}CH(CH_2)_7$
Octadecatrienoic		$COOH$ (all *cis*)
Eicosenoic	20	$CH_3(CH_2)_7CH{=}CH(CH_2)_9$
cis-11-Eicosenoic		$COOH$ (*cis*)

saponification. During the process of extracting olive oil from the fruit, hydrolysis reactions can occur to release free fatty acids. The existence of free fatty acids yields a higher acid content and a lower quality oil.

The composition of an oil is usually expressed in terms of the percentages of the various acids obtained from its saponification. Fatty acids fall into two main categories: saturated and unsaturated. Saturated fatty acids contain all single bonds between neighboring carbon atoms. Unsaturated fatty acids contain one or more carbon–carbon double bonds. Unbranched, straight-chain molecules with an even number of carbon atoms are dominant among the saturated fatty acids. Unsaturated fatty acids generally contain *cis* configurations about the carbon–carbon double bonds. Olive oils contain the saturated fatty acids myristic acid, palmitic acid, and stearic acid, and the unsaturated fatty acids oleic acid, linoleic acid, and linolenic of varying composition (Table 13.2).

In particular, olive oil contains a high percentage of oleic acid in comparison to the oils of other plants (see Table 13.3).

One way to determine the fatty acid composition of a plant oil is by the technique of high-performance liquid chromatography. This technique is a separation

TABLE 13.3 ► Percent Fatty Acid Composition of Common Oils from Plants[4,5]

Oil	Myristic saturated C-14	Palmitic saturated C-16	Stearic Acid saturated C-18	Oleic mono-unsaturated C-18	Polyun-saturated C-18	Other
Coconut	18	11	2	8	—	61
Corn	1	10	3	50	34	2
Cottonseed	1	23	1	23	48	4
Linseed	—	6	3	19	72	—
Olive	—	7	2	85	5	1
Palm	1	40	6	43	10	—
Palm kernel	14	9	1	18	1	57
Peanut	—	8	3	56	26	7
Safflower	—	4	3	17	76	—
Sesame	—	9	4	45	40	2
Soybean	—	10	2	29	57	2
Sunflower	—	6	2	25	66	1
Wheat germ	—	13	4	19	62	2

method useful for analyzing complex mixtures. The plant oil containing a mixture of triglycerides is initially dissolved in a liquid solvent. HPLC involves using high pressure to force this liquid "mobile phase" carrying the solutes to be separated through a column packed with solid particles with a high surface area. The particles, often silica, are usually coated with an organic film that is chemically or physically bonded to the surface. Successful analysis of a mixture requires a proper balance of the intermolecular forces between the three participants in the separation process — the solute (here, the triglycerides), the solvent used as the mobile phase, and the column particles composing the "stationary phase." As the mobile phase is passed through the column, the triglycerides interact with the stationary phase to varying degrees. Some triglycerides will interact strongly with the column material, increasing the length of time that these solutes spend in the column. Other triglycerides will exhibit a weaker interaction and move through the column quickly. The time that a particular solute spends in the column is known as the *retention time*. For a given set of experimental conditions (choice of solvent and column-packing material, temperature, flow pressure, etc.), a chemical substance will have a characteristic retention time. Thus, qualitative analysis of a sample is possible by analyzing retention times. By using an additional analytical technique such as UV/visible absorbance, a chromatogram is produced with time plotted on the *x* axis and absorbance at a specified wavelength plotted on the *y* axis. A given plant oil will yield a characteristic chromatogram that can be used as a signature of the sample. The area of a peak on the chromatogram is a measure of the concentration of the triglyceride in the sample.

KEY TERMS:	triglycerides	esters	alcohol
	carboxylic acid	hydrolysis	chromatography

References

[1] "Tasting Olive Oil." Michael Burr's Olive's Site,
http://www.senet.com.au/~betaburr/Tasting.html#Trade

[2] "The Olive Oil Source."
http://www.oliveoilsource.com/mill_and_press_facts.htm#Cleaning

[3] "Olive Oil Definitions." The Olive Oil Source,
http://www.oliveoilsource.com/definitions.htm

[4] "The Quality of Olive Oil Chemistry." http://www.oliocarli.com/htm/equal2.htm

[5] Carl H. Snyder, *The Extraordinary Chemistry of Ordinary Things*, 2nd ed. (New York: Wiely, 1995).

13.17 | How Can Sucralose, an Artificial Sweetener Made from Sugar, Contain *No* Calories?

During digestion, complex carbohydrates (which are composed of multiple individual sugar molecules known as polysaccharides) are cleaved into single molecule units, absorbed, and transported to the cells in the blood. These sugar molecules are either metabolized immediately to provide energy for the cell or stored in liver and muscle cells as glycogen to be used for future energy needs. If sucralose is made from sugar, how can the sweetener have no caloric value?

The Chemical Basics

Did you know the average American consumes the equivalent of 20 teaspoons of sugar each day?[1] The "non-nutritive" sweetener industry is described as a billion-dollar industry with projections of even more rapid expansion in the next few years. What do chemists look for in their search for an "ideal" sweetener? Consumers seek good-tasting, nontoxic, low-caloric sweeteners. Chemists in the sweetener industry add further demands: an inexpensive, easy-to-synthesize product that is readily soluble in water and resists degradation by heat and light is of prime importance. The chemical structure of sucralose keeps the sweetener intact as it passes through the acidic environment of the stomach. Thus, sucralose is not metabolized and passes through the body unchanged and is eliminated, adding no calories. The stable chemical structure of sucralose also contributes to its exceptional heat stability and long shelf-life. Sucralose's stability and sweetness, coupled with its lack of an unpleasant aftertaste , account for the wide popularity of sucralose in a variety of foods and beverages.[2]

The Chemical Details

Sucralose, also known as Splenda, is derived from cane sugar (sucrose) and is manufactured in a five-step process[3,4] by McNeil Speciality Products, a subsidiary of Johnson & Johnson, of New Brunswick, NJ. Sucrose is known as a disaccharide composed of one molecule of glucose and one molecule of fructose. In the sucralose manufacturing process, the sucrose sugar molecule is modified by selectively substituting three atoms of chlorine for three hydroxyl (–OH) groups on the sugar molecule. Asterisks (*) in Fig. 13.17.1 indicate the three –OH groups

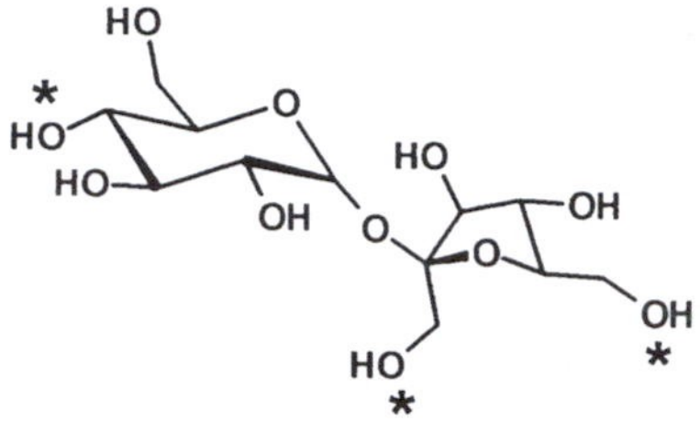

Figure 13.17.1 ► The molecular structure of sucrose, a disaccharide of glucose and fructose. Asterisks indicate the hydroxyl groups replaced with chlorine atoms in the synthetic sweetner.

replaced by a chlorine atom in sucralose.

As a consequence of adding the chlorine atoms, the sucralose molecule becomes essentially chemically inert and is not broken down by gastric juice into its component parts — derivatives of glucose and fructose. In other words, sucralose is not metabolized and passes through the body unchanged and is eliminated, adding no calories. Some hydrolysis of sucralose may occur under conditions of prolonged storage at elevated temperatures in highly acidic aqueous food products. What are the products of such a decomposition reaction? The hydrolysis decomposes the ether into two alcohols, the monosaccharides 4-chloro-4-deoxy-galactose (4-CG) and 1,6-dichloro-1,6-dideoxyfructose (1,6-DCF).[3]

KEY TERMS: **hydrolysis** **hydroxyl group** **ether** **alcohol**

References

[1] "Sugar Substitutes: Americans Opt for Sweetness and Lite." John Henkel, FDA Consumer Magazine, November–December 1999, http://www.fda.gov/fdac/features/1999/699_sugar.html

[2] "Low-Calories Sweeteners: Sucralose." http://www.caloriecontrol.org/sucralos.html

[3] "Food Additives Permitted for Direct Addition to Food for Human Consumption; Sucralose." Federal Register: August 12, 1999 (Vol. 64, No. 155), http://class.fst.ohio-state.edu/fst621/Lectures/neotame.htm

[4] "Sucralose — A High Intensity, Noncaloric Sweetener." PEP Review 90-1-4, Process Economics Program, SRI Consulting, http://pep.sric.sri.com/Public/Reports/Phase_90/RW90-1-4/RW90-1-4.html

Related Web Sites

▶ "FDA Approves New High-Intensity Sweetener Sucralose." FDA Talk Paper T98-16, April 1, 1998, *The Diabetes Monitor*, http://www.diabetesmonitor.com/sucralos.htm

▶ "Digestion." Chemistry 431 Biochemistry, Richard A. Paselk, Department of Chemistry, Humboldt State University, http://www.humboldt.edu/~rap1/C431.F99/C431Notes99/C431n22oct.htm

Other Questions to Consider

| 7.7 | How does a bullet-proof vest work? How is it made? *See* p. 92. |

| 12.6 | What is the purpose of the thread that runs vertically through the clear field on the face side of U.S. currency? *See* p. 152. |

| 13.3 | How do forensic chemists use visible stains to trap thieves? *See* p. 170. |

| 13.4 | Why is willow bark known as "Nature's aspirin"? *See* p. 172. |

| 13.6 | How do "sniffing dogs" detect narcotics or explosives? *See* p. 176. |

Chapter 14 ▶

Connections to Polymers

14.1   How Does a Timed-Release Medicine Work?

The formulation of timed-release medicines is based on the specific response of polymeric coatings to their chemical environment. The chemical packaging of these medicines determines the precise conditions for effective control and sustained dosage of these drugs.

The Chemical Basics

Physicians utilize a variety of protocols and therapies to heal and cure patients. Some medical treatments require the sustained application of a drug for maximum effectiveness. When hospitalized, a patient can receive this continued drug delivery through an intravenous (IV) unit. Some oral medications are chemically formulated to achieve this same effect. A medicine taken orally can often be gradually released in the body over a specified time interval by carefully designing the coating that encapsulates the medication. For example, the decongestant Contac contains numerous tiny beads of medicine that are covered by a water-soluble polymeric coating of varying thickness. The thicker the coating, the longer the time required for the coating to dissolve in water and the slower the release of the medicine. The claim of an effective "twelve-hour medicine" is based on the precise combination of beads of medicine with prescribed thicknesses of polymeric coating to sustain the controlled release for an extended period of time. Thin coatings obviously dissolve quickly, while thicker layers take longer (up to 12 hours, in this case). Timed-release formulations using "microencapsulation technology" are also used for agricultural purposes (e.g., fertilizers) and insect control (e.g., a six-month pest control).

The Chemical Details

FMC designs one such water-soluble coating for timed-release medications — Aquacoat ECD, a 30% by weight aqueous dispersion of ethylcellulose. This polymer is used to coat drug-layered beads that are delivered using gelatin capsules for a pH-independent sustained release. Cellulose is a natural polymer containing repeating glucose units (monomers). Cellulose forms during a *condensation polymerization reaction* (Fig. 14.1.1) in which each new link of glucose monomers

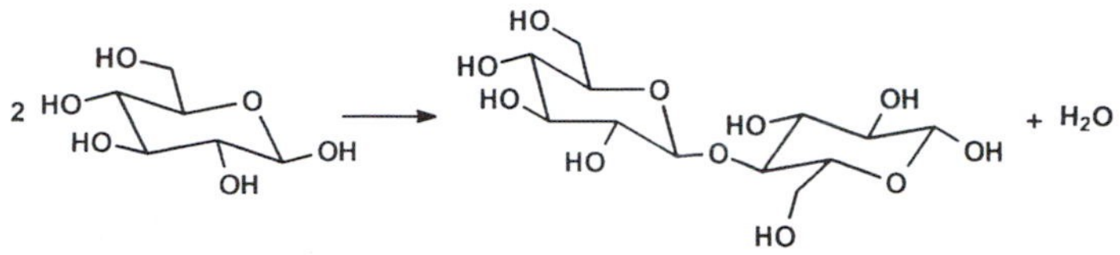

Figure 14.1.1 ► The formation of cellulose during a condensation polymerization reaction in which each new link of glucose monomers releases a water molecule.

releases a water molecule. The –OH and –CH_2OH substituents on the glucose rings are replaced with –OCH_2CH_3 and –$CH_2OCH_2CH_3$ groups in ethylcellulose (Fig. 14.1.2).

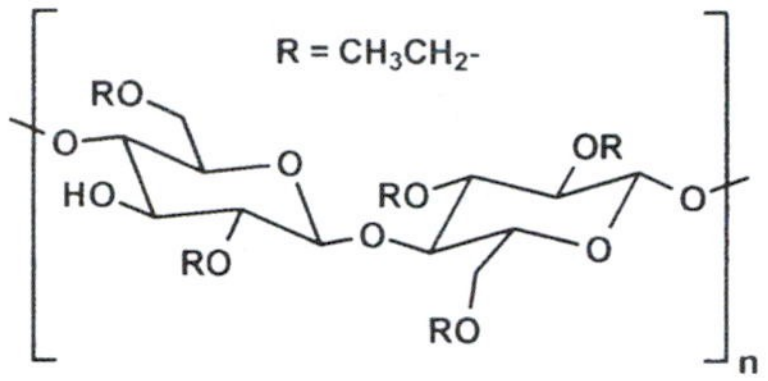

Figure 14.1.2 ► The repeating unit of the polymer ethylcellulose.

Some extended-release preparations are designed with a coating that responds to the acidity of its environment. The polymeric coating of the medicine is formulated for stability during oral delivery and for eventual solubility at the intended organ. The contrasting acidic content of the stomach and the more basic environment of the intestines enable these formulations to function. For example, hydroxypropyl methylcellulose phthalate (HPMCP) (Fig. 14.1.3) is an *enteric*

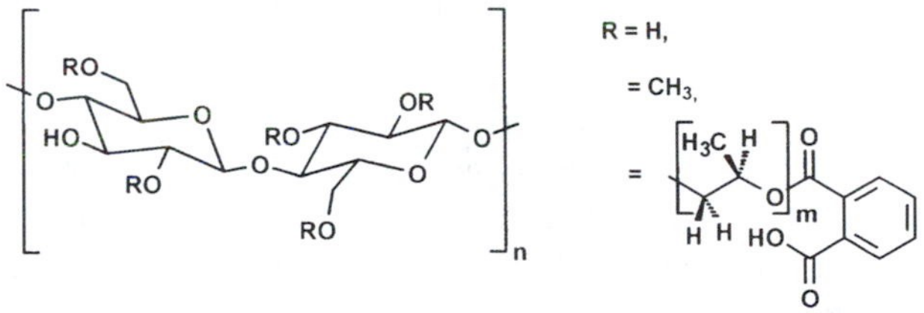

Figure 14.1.3 ► The repeating unit of the polymer hydroxypropyl methylcellulose phthalate (HPMCP).

(i.e., solubilized in the intestinal tract) coating designed to protect acid-sensitive drugs from being destroyed by gastric acid in the stomach. In a more alkaline environment, deprotonation of the –COOH carboxyl groups (to form –COO– carboxylate functionalities) is believed to enable the dissolution of the polymeric HPMCP, thereby releasing the encapsulated drug.[1] Enteric coating materials are specifically used to target timed-release medications to treat colon inflammations or other disorders of the digestive tract. In addition, enteric coatings are placed on aspirin caplets designed for the temporary relief of arthritic and rheumatic pain, muscle aches, joint pain, and back pain. The coating allows the caplet to pass through the stomach to the intestine before it dissolves, to help prevent stomach irritation.

Polymer coatings responsive to temperature or moisture form the basis for medications delivered transdermally using patches on the skin or internally via inserts implanted in the body. Oral nitroglycerin (Fig. 14.1.4) tablets to prevent

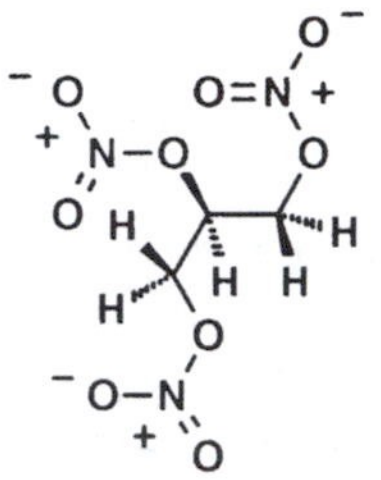

Figure 14.1.4 ▶ The molecular structure of nitroglycerin.

angina attacks or scopolamine (Fig. 14.1.5) to protect against motion sickness are

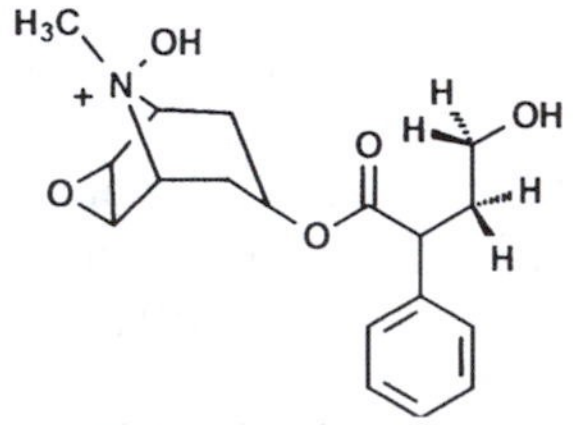

Figure 14.1.5 ▶ The molecular structure of scopolamine.

two examples of drugs that can penetrate the skin and flow into the bloodstream at a rate dictated by a rate-controlling membrane in the patch. Hydrolysis of nitroglycerin leads to the formation of the reactive free radical nitric oxide, NO. NO activates guanylate cyclase to produce cyclic guanosine monophosphate (GMP); cyclic GMP decreases cellular calcium levels, thereby causing dilation or expansion of the blood vessels to reduce myocardial oxygen demand.[2] Dilation of the arteries increases blood flow to the heart and relieves the chest pains that result from an insufficient supply of oxygen to the heart muscle. The sophistication of the technology of timed-release medications arises from the extensive structural control that the chemist has available for the design of polymeric coatings and matrices.

> **KEY TERMS:** polymer microencapsulation enteric

References

[1] G. H. W. Sanders, J. Booth, and R. G. Compton, "Quantitative Rate Measurement of the Hydroxide Driven Dissolution of an Enteric Drug Coating Using Atomic Force Microscopy." *Langmuir* **13** (1997), 3080–3083.

[2] J. Ahlner, R. G. G. Andersson, K. Torfgard, and K. L. Axelsson, "Organic Nitrate Esters: Clinical Use and Mechanisms of Actions." *Pharmacol. Rev.* **43** (1991), 351–423.

Related Web Sites

▶ "Aquacoat ECD." FMC Corporation, Pharmaceutical Division, Products, http://www.avicel.com/products/ec.html

▶ "Transdermal Drug Delivery Systems." PHA 5110 Pharmaceutics II, University of Florida, College of Pharmacy, Department of Pharmaceutics, http://www.cop.ufl.edu/safezone/prokai/pha5110/tdds.htm

▶ "Patches, Pumps and Timed Release: New Ways to Deliver Drugs." Marian Segal, U.S. Food and Drug Administration, http://www.fda.gov/bbs/topics/CONSUMER/CON00112.html

▶ "Drug Patches Catch On." Lalitha Gopinath, Global Bytes: Chemistry in Britain, September 1997, http://www.chemsoc.org/chembytes/ezine/1997/patches.htm

14.2 How Does "Scratch & Sniff" and Carbonless Copy Paper Work?

The clever combination of chemistry and microsphere technology enables advertisers to take advantage of the incredible marketing power of scent. These innovations have also modernized paper-based recordkeeping and business transactions.

The Chemical Basics

These two seemingly dissimilar applications have a common basis — both are examples of the *pressure-sensitive* release of a chemical. How are these products designed? Tiny spherical capsules (*microcapsules* or *microspheres*) with a glass or polymer shell are filled with a liquid core and glued onto paper. For a "scratch-and-sniff" ad, the core of the microcapsules contains a liquid with the desired scent; for carbonless paper, a liquid ink or dye is encapsulated within the microspheres that adhere to the underside of the paper. When the paper ad is scratched, the shell of the microsphere is broken. By exposing the liquid cavity, the scent of the enclosed perfume or fragrance is easily detected. Similarly, when pressure is applied to the copy paper through a pen or typewriter, the encapsulated colorless dye is released and reacts with an acidic chemical on the paper underneath to darken and transfer ink to the underlying paper.

Microencapsulation is the term for the shielding of solid, liquid, or gaseous active ingredients by enclosure of the active component within a protective shell. To release the contents of the microcapsule, four techniques are often employed. Scratch-and-sniff cards and carbonless copy paper rely on the *mechanical rupture* of the microsphere shell induced by pressure. Timed-release medicines and agrochemicals are often based on *water-soluble polymeric coatings* that dissolve over time. (See Question 14.1.) Sustained release of medications is also achieved via *diffusion* of the active components through microcapsule walls. *Thermal breakdown* or *melting* of a solid microcapsule wall by increased temperature is also a release mechanism, used to liberate fat-encapsulated baking soda in packaged baking mixes, for example.

The Chemical Details

The polymer and glass microspheres employed in the pressure-sensitive release of chemicals range in size from 1 μm to 1 mm in diameter. (For comparison, a human hair is typically 80–100 μm in diameter.)[1] A scanning electron micrograph illustrating the morphology of the particles appears in color Fig. 14.2.1.[2]

A variety of techniques are employed to achieve microencapsulation. One of the earliest methods, developed in the 1930s, is known as *coacervation* or *phase separation*. This method reproducibly applies a uniform, thin, polymeric coating to small particles of solids or to droplets of pure liquids and solutions. To apply this method, four components are necessary: the material to be coated (core), a wall-forming polymer, a suitable solvent (liquid-manufacturing vehicle), and a coacervation (phase-separation) inducer.[3] For water-soluble core materials, a polymer soluble in a nonpolar solvent such as cyclohexane must be chosen (e.g., ethylcellulose). For core substances not miscible in water, a water-soluble polymer (e.g., gelatin) is required. The polymer must also be chemically compatible with the core material and be capable of forming a cohesive film on the core surface. Selection of an appropriate polymer, whether natural or synthetic, should also address the needs for a coating material with the requisite strength, flexibility, impermeability, and stability.

In the coacervation process, the core substance is first added to a homogeneous solution of the selected solvent and polymer. Mechanical agitation is used to disperse the immiscible core to create tiny droplets suspended in solution (i.e., an emulsion). The coacervation or phase separation phenomenon is then induced by several means, such as changing the temperature and/or acidity of the polymer solution or adding salts, nonsolvents, or incompatible (immiscible) polymers to the polymer solution. For example, by adjusting the acidity of the system through the addition of an acid, a phase separation may be induced. In other words, two immiscible liquid phases are created with different amounts of the solubilized polymer in each phase. The supernatant phase has low polymer concentrations, while the coacervate phase has a relatively high concentration of the polymers. The polymer is selected so that the coacervate phase preferentially adsorbs onto the surface of the dispersed droplets to form the shell of the microcapsules. Once the fluid film of polymer is deposited from the coacervate onto the core, the shell must be solidified. Cooling and further chemical reaction with a cross-linking agent such as formaldehyde hardens the microcapsule walls. The microcapsules are then separated by settling or filtration, and then washed, filtered, and dried.

KEY TERMS: **microencapsulation**   **microsphere**

References

[1] "Which Object is Closest in Length to a Micrometer?" "The Advanced Light Source: A Tool for Solving the Mysteries of Materials." Lawrence Berkeley National Laboratory, http://www.lbl.gov/MicroWorlds/ALSTool/micrometer.html

[2] "Scratch-and-Sniff Paper." The Museum of Science, Boston, http://www.mos.org/sln/sem/scratch.html

[3] "Taste-Masking/Quick Dissolving." Eurand Research and Development, http://www.eurand.com/wt/tert.php3?page_name=taste_masking

Related Web Sites

▶ "Preparation and Characterisation of Novel pH Responsive Microgel Particles." Matt Hearn, Department of Chemistry, University of Bristol, http://www.chm.bris.ac.uk/vincent/matt.html

▶ "The Chemistry behind Carbonless Copy Paper," Mary Anne White, *J. Chem. Educ.* **75** (1998), 1119, http://jchemed.chem.wisc.edu/Journal/Issues/1998/sep/abs1119 .html

▶ "The Art and Science of Microencapsulation." J. Franjione and N. Vasishta, *Technology Today*, http://www.swri.org/3pubs/ttoday/summer/microeng.htm

▶ Microencapsulation, Faculty of Pharmaceutical Sciences, Chulalongkorn University, http://www.pharm.chula.ac.th/microcap/MIC.HTM

▶ "Review of Pharmaceutical Controlled Release Methods and Devices." Paul A. Steward, http://www.initium.demon.co.uk/rel_nf.htm#Drug_Deliv

14.3   How Do "Repositional Self-Adhesive Paper Notepad Sheets" (Post-it Notes) Work?

Post-It Notes are ubiquitous forms of communication at the office, in schools, and at home. This extremely useful invention is actually the result of a failed experiment. How does chemistry contribute to the function of this convenient product?

The Chemical Basics

Office supplies and office communication have been revolutionized by the introduction of the 3M product Post-it Notes in 1980. The temporary adhesive of these self-sticking notes actually was initially rejected as a glue by its inventor, 3M chemist Spencer Silver, because of its impermanence. However, 3M researcher Art Fry found a niche for the adhesive — as an adhesive for a temporary bookmark for his choir hymnal.[1] Key to the performance of these removable, repositional adhesive products is the application of the adhesive to the backing of a note sheet via tiny microspheres rather than a continuous film. With an average particle diameter of 25–45 μm, the microsphere adhesives form a discontinuous layer that assists in retaining the ability to re-apply the note to new surfaces. Traditional adhesive tape contains particles of smaller dimension (typically 0.1 to 2.0 μm) that coalesce to form a continuous film that limits removal.[2]

The Chemical Details

The tacky polymeric microspheres that comprise the pressure-sensitive adhesive layers of "repositionable notes" are patented inventions. One such material (U.S. Patent 5,714,237)[3] is prepared by a *free-radical polymerization reaction* of isooctyl acrylate (Fig. 14.3.1) in the presence of polyacrylic acid with a *chain-*

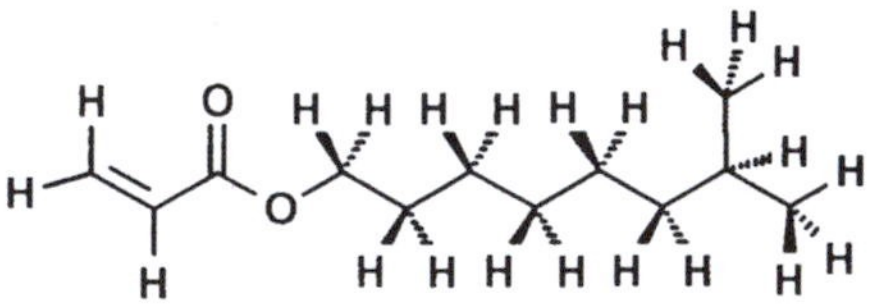

Figure   14.3.1 ▶   The molecular structure of isooctyl acrylate, a monomer in a free-radical polymerization reaction.

transfer agent (dodecanethiol) (Fig. 14.3.2), an *initiator* (monomer soluble bis-(t-butylcyclohexyl)peroxycarbonate) (Fig. 14.3.3), and a detergent (ammonium lauryl sulfate) (Fig. 14.3.4). The adhesive polymeric composition is recovered and blended with a solid coating mixture to create the microspheres.

Free-radical polymerization reactions are also known as *chain-growth* or *addition* polymerization reactions. Let's look at a chain-growth polymerization re-

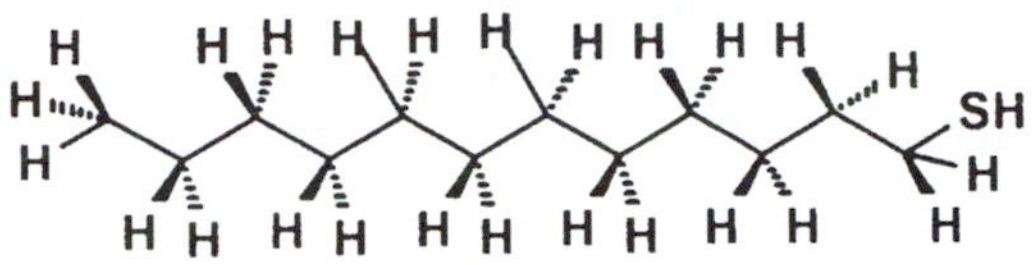

Figure 14.3.2 ▶ The molecular structure of dodecanethiol, a chain-transfer agent in a free-radical polymerization reaction.

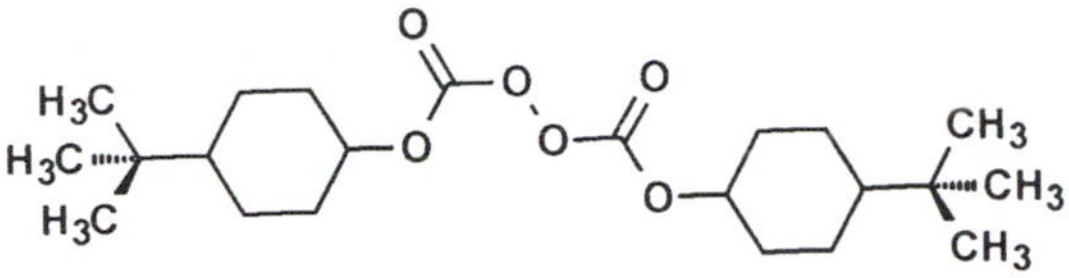

Figure 14.3.3 ▶ The molecular structure of di-(4-tert-butylcyclohexyl)peroxydicarbonate, an initiator in a free-radical polymerization reaction.

actions in more detail. As the name suggests, these processes convert small monomer molecules to larger polymers by the successive addition of monomer molecules onto the reactive ends of a growing polymer. An initiator is required in chain-growth polymerization reactions. The function of the initiator is to react with the monomer to form another reactive compound, thereby beginning (initiating) the linking process. Under the proper experimental conditions, the peroxycarbonate initiator easily cleaves at its oxygen–oxygen bond to create unstable species called free radicals (or simply radical species) that are characterized by an unpaired electron:

$$C_6H_{11}OC(O)OOC(CH_3)_3 \longrightarrow C_6H_{11}OC(O)O \cdot + \cdot OC(CH_3)_3.$$

One of these free radicals may decompose further by releasing carbon dioxide and generating a new free radical.

$$C_6H_{11}OC(O)O \cdot \longrightarrow C_6H_{11}O \cdot + CO_2 \text{ (g)}.$$

These reactive species initiate the linking process by adding to the double bond of the chosen acrylate monomer (isooctyl acrylate) to create another reactive radical species. This linking process may be represented by

$$I \cdot + H_2C=CHX \longrightarrow \cdot H_2C-CHI \cdot .$$

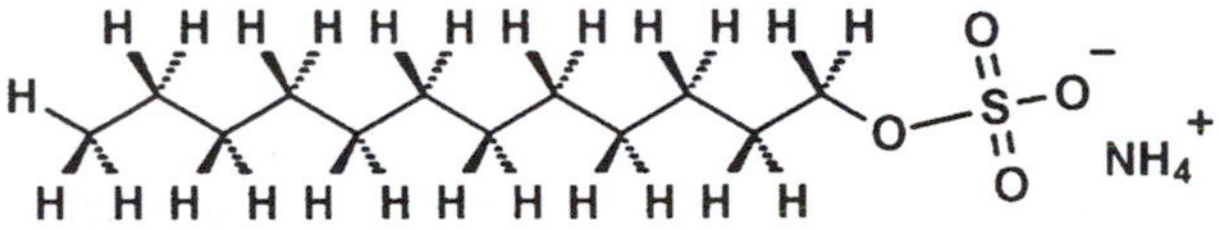

Figure 14.3.4 ▶ The molecular structure of ammonium lauryl sulfate, a detergent.

This new free radical adds to the double bond of another monomer molecule, growing the polymer chain. The polymerization process ends as the unpaired electrons of two free radicals combine to form a single bond:

$$R\text{-}CH_2\text{–}CHX \cdot + \cdot CHX\text{–}CH_2\text{-}R \longrightarrow R\text{–}CH_2\text{–}CHX\text{–}CHX\text{–}CH_2\text{–}R.$$

KEY TERMS: monomer   polymer   free-radical* polymerization

                        free radical   initiator   *or *chain growth*, *addition*

References

[1] "Post-it Notes Become a Best-Selling Office Product." 3M, http://mustang.coled.umn.edu/inventing/Post3m.html

[2] "Microsphere Technology." Advance Polymers International, http://www.msphere.com/micros.htm

[3] "Patent Watch." *Chemtech* (April 1998), 25.

Related Web Sites

- "3M Chemist Wins Creative Invention Award." http://www.mmm.com/front/silver/index.html

- Polymer Synthesis, Polymers and Liquid Crystals, Virtual Textbook, Case Western University, http://plc.cwru.edu/tutorial/enhanced/files/polymers/synth/s ynth.htm

- "Free Radical Vinyl Polymerization." Department of Polymer Science, University of Southern Mississippi, http://www.psrc.usm.edu/macrog/radical.htm

- "Fundamental of Polymer for Fiber Applications." American Fiber Manufacturers Association, http://www.fibersource.com/f-tutor/poly.htm

14.4   What Is "Shatterproof" Glass?

How does chemistry increase the resistance of glass to impact and shattering?

The Chemical Basics

The shatterproof glass used in impact-resistant windows is actually not a glass material derived from silicon dioxide. Instead, shatterproof glass is a *thermoset plastic* or thermoplastic, i.e., a pliable material that is even easier to mold when hot. Shatterproof windows are made using a specific thermoset material known as *polycarbonate* of bisphenol A (or bisphenol A polycarbonate). This clear,

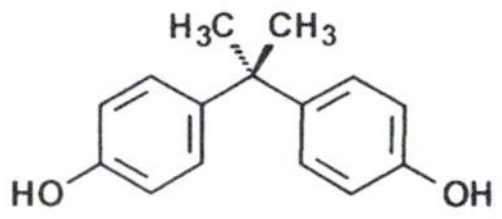

Figure 14.4.1 ▶ The repeating units of the monomer bisphenol A used to construct the polymeric thermoset material used in shatterproof glass.

glassy polymer is constructed from repeating units of the monomer bisphenol A (Fig. 14.4.1). Thus, bisphenol A polycarbonate is also considered an example of a *heterochain polymer*, containing atoms in addition to carbon in its backbone chain. Bisphenol A polycarbonate is also a member of the larger polymer family of *polyesters*, i.e., polymers formed from a large number of smaller molecules, or monomers, by establishment of ester linkages (Fig. 14.4.2) between them.

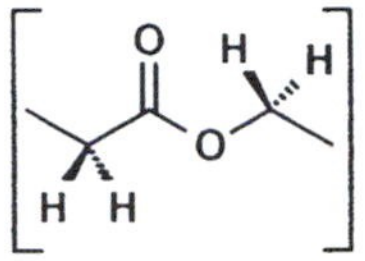

Figure 14.4.2 ▶ The ester linkage found in the polymers known as polyesters.

General Electric markets the polycarbonate of bisphenol A as Lexan.[1] Similar bisphenol A polycarbonate sheets are marketed by Rohm and Haas as the product Tuffak.[2]

The Chemical Details

Polycarbonate is a generic term for the class of polymers consisting of long-chain linear polyesters of carbonic acid, H_2CO_3, and aromatic alcohols known as *phenols* that possess two hydroxyl groups.

The synthesis of polycarbonate of bisphenol A begins with the reaction of bisphenol A and sodium hydroxide to obtain the sodium salt of bisphenol A, as in Fig. 14.4.3. The sodium salt of bisphenol A is then reacted with phosgene to

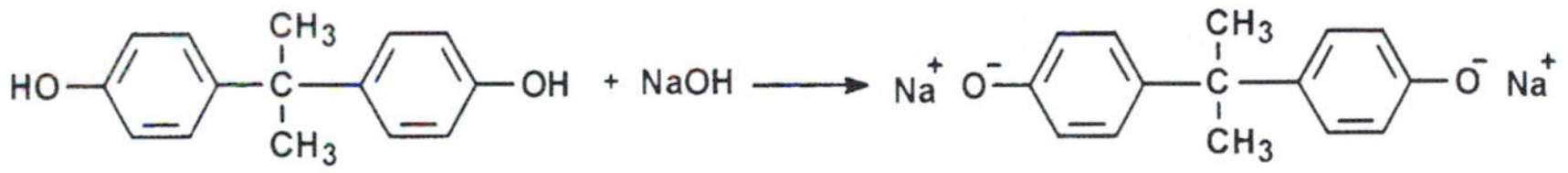

Figure 14.4.3 ▶ The reaction of bisphenol A and sodium hydroxide to obtain the sodium salt of bisphenol A.

produce the polycarbonate, as diagrammed in Fig. 14.4.4.[3] Alternative synthesis procedures under exploration replace the solvent phosgene with carbon dioxide to accomplish the final step of the synthesis.[4]

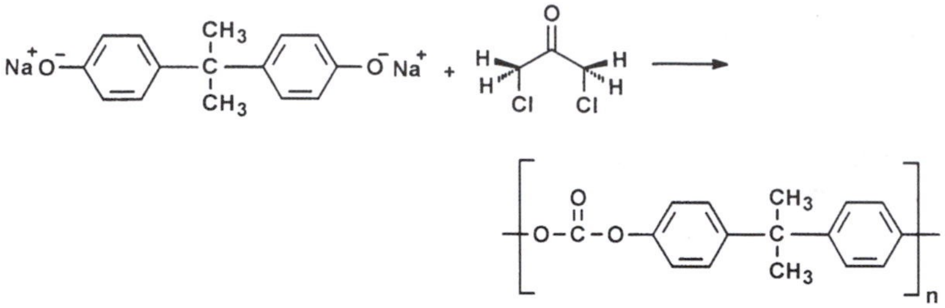

Figure 14.4.4 ▶   The reaction of the sodium salt of bisphenol A with phosgene to produce the polycarbonate.

The properties of the polycarbonate of bisphenol A are directly related to the structure of the polymer. The molecular stiffness associated with this polycarbonate arises from the presence of the rigid phenyl groups on the molecular chain or backbone of the polymer and the additional presence of two methyl side groups. The transparency of the material arises from the amorphous (noncrystalline) nature of the polymer. A significant crystalline structure is not observed in the polycarbonate of bisphenol A because intermolecular attractions between phenyl groups of neighboring polymer chains in the melt lead to a lack of flexibility of the chains that deters the development of a crystalline structure.

Another polymer used for unbreakable windows is poly(methyl methacrylate). PMMA is a vinyl polymer, made by free radical vinyl polymerization from the monomer methyl methacrylate, according to the reaction in Fig. 14.4.5.[3] Rohm and Haas introduced this PMMA-based shatterproof glass as Plexiglas.[5]

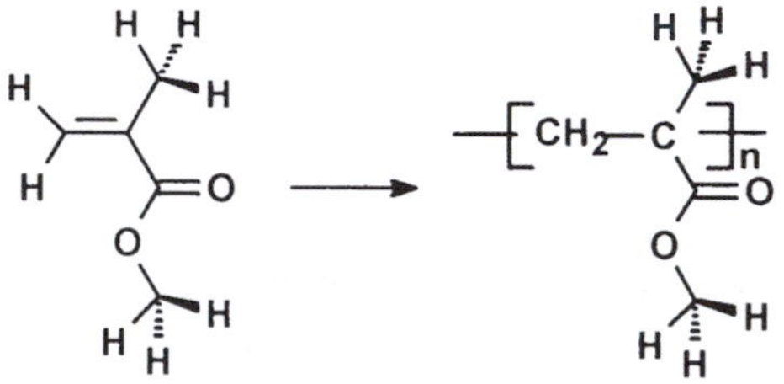

Figure 14.4.5 ▶   Free-radical vinyl polymerization of the monomer methyl methacrylate to form the vinyl polymer poly(methyl methacrylate) (PMMA).

Imperial Chemical Industries also produces PMMA as Lucite.

KEY TERMS: polycarbonate polyester   polymer   thermoplastic
                       monomer   thermoset   plastic   heterochain polymer

References

[1] "Material Safety Data Sheet, Lexan Resin." Plastixs, LLC,
http://www.plastixs.com/pdfs/msds/41412N.pdf

[2] Material Safety Data Sheet, TUFFAK Polycarbonate Sheet, Atofina Chemicals, Inc.,
http://www.atofinachemicals.com/msds/426.pdf

[3] "Polycarbonates." Department of Polymer Science, University of Southern Mississippi.
http://www.psrc.usm.edu/macrog/pc.htm

[4] "No. 30 Environmentally Benign Synthesis of Aromatic Polycarbonates." Summary of
Lectures and Posters, http://www.gscn.net/event/meeting/summary.html

[5] "Material Safety Data Sheet Plexiglas Acrylic Sheet." Atofina, Inc.,
http://www.atofinachemicals.com/msds/421.pdf

Related Web Sites

▶ "Making Polycarbonate." Department of Polymer Science, University of Southern Mississippi, http://www.psrc.usm.edu/macrog/pcsyn.htm

14.5 | Why Does Superglue Stick to Almost Every Surface?

One commercial adhesive is marketed with the following claims: "High Strength Adhesive;" "Durable Bonding;" "Fast Acting;" "Bonds Metals, Rubber, Ceramics, Plastics, Glass, Wood, Veneers, Fabrics, Vinyl, Cardboard, Cork, Leather, Nylon, and Other Similar Surfaces."[1] How can one substance act as a general purpose adhesive with affinity for so many types of surfaces?

The Chemical Basics

One class of adhesives known as *superglues* consists of synthetic organic polymers that provide strong and rapid adhesion. These adhesives are unusual in that the polymerization process to form the adhesive occurs upon exposure of the monomer to water. Under most conditions, atmospheric moisture is sufficient to form a strong adhesive. Superglues stick to a variety of surfaces since a film of moisture exists on almost any surface. The quality of bonding will vary with the humidity; the higher the humidity, the better the set.

The Chemical Details

The adhesive marketed under the tradename "Superglue" contains the monomer methyl α-cyanoacrylate (Fig. 14.5.1). A variety of cyanoacrylates are commercially sold as contact adhesives with the alkyl group –R denoted in Fig. 14.5.2 varying from a methyl group to produce ethyl, isopropyl, allyl, butyl, isobutyl,

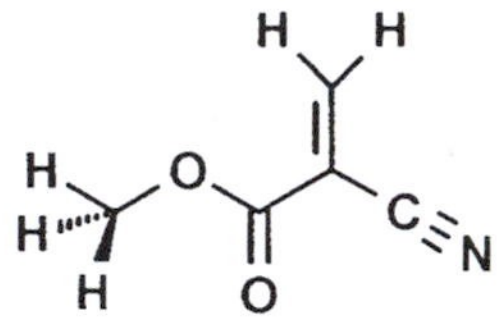

Figure 14.5.1 ▶   The monomer methyl α-cyanoacrylate found in the adhesive marketed under the trademark "Superglue."

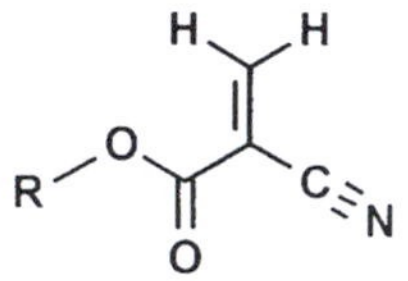

Figure 14.5.2 ▶   The general class of cyanoacrylates with varying alkyl group –R, sold as contact adhesives.

methoxyethyl, and ethoxyethyl cyanoacrylate esters.[1–3] The properties of the adhesive (e.g., setting time, strength, durability) will vary with the substitution. All of these monomers undergo polymerization in the presence of water. In fact, water serves as an initiator for the polymerization according to an anionic vinyl polymerization mechanism.[4] The overall polymerization process for forming the cyanoacrylate polymer is represented schematically by the reaction in Fig. 14.5.3.

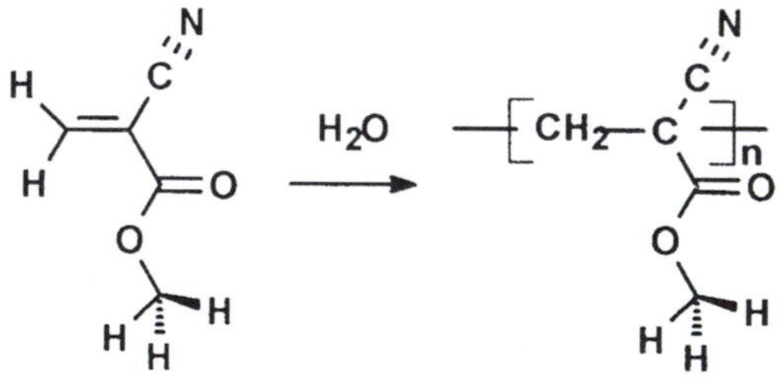

Figure   14.5.3 ▶   The overall polymerization process for forming the cyanoacrylate polymer.

While the formation and subsequent cross-linking of the polymer is one factor in the effectiveness of an adhesive, the adhesive strength of the polymer–surface interface is also critical. Both physical and chemical considerations influence the bonding. A rough or porous surface is generally more effective for "locking" an adhesive to a substrate. However, "chemically active" sites on the surface for promoting interaction via hydrogen bonding, strong dipole–dipole interactions, or other intermolecular attractions also contribute to the adhesion properties.

KEY TERMS: monomer polymer adhesion

References

[1] Tech Hold Cyanoacrylate. http://www.techspray.com/2502info.htm

[2] Polycyanoacrylates, Department of Polymer Science, University of Southern Mississippi. http://www.psrc.usm.edu/macrog/pca.htm

[3] Technical Data Sheet for Gel Set 44 Ethyl Cyanoacrylate, http://www.holdtite.com/english/technical/tech/ca44.htm

[4] Anionic Vinyl Polymerization, Department of Polymer Science, University of Southern Mississippi. http://www.psrc.usm.edu/macrog/anionic.htm

Related Web Sites

► "Reactive Adhesives, Proactive Chemistry." http://www.chemsoc.org/chembytes/ezine/1998/glue.htm

► "Superglue." Bruce Sterling, http://www.eff.org/pub/Publications/Bruce_Sterling/FSF_columns/fsf.07

14.6 What Is the Difference between Hard and Soft Contact Lenses?

The number of options for contact lens wearers has been transformed by innovative advances in the chemistry of the polymeric materials used to create these vision products.

The Chemical Basics

Most everyone is familiar with the contrasting resilient, pliable nature of soft contact lenses and the brittle character of hard lenses. Both lenses are constructed from polymers, but the differing chemical composition of each polymer leads to considerably different physical properties.

Hard contact lenses are composed of a polymer that repels water because the constituent repeating units (the *monomers* that link together to form the polymer) are nonpolar, *hydrophobic* segments. The first hard contact lens was constructed in 1948 from the monomer known as methyl methacrylate (MMA), yielding the polymer poly(methyl methacrylate) or PMMA. This material offers durability, optical transparency, and acceptable wettability for optimal comfort. Today the rigid lens material of hard contact lenses is often constructed by combining MMA with one or more additional hydrophobic monomers to provide better gas permeability.

The first soft contact lenses were also constructed with a polymeric material containing a single monomeric unit. The added pliability of the soft lens was derived from the more *hydrophilic* nature of the monomer, enhancing the ability of the polymer to absorb water and provide greater comfort to the lens wearer. This monomer is a derivative of MMA known as hydroxyethyl methacrylate (HEMA). A number of hydrophilic monomers are used in soft lenses today; these materials are referred to as *hydrogels* because of their ability to absorb significant amounts of water yet remain insoluble.

The soft extended wear lenses popular today are composed of polymers with more than one type of repeating unit, i.e., *copolymers*. For extended wear a lens with greater oxygen permeability is needed, for the cornea relies on direct oxygen transmission from the atmosphere as a consequence of the lack of blood vessels within the corneal framework. Scientists have found that the higher the water content of a hydrogel polymer, the more extensive the oxygen permeability of the lens formulated from that polymer. A 70% water content is desirable for extended periods of lens wear, expressed as a percentage of the total weight of the polymer. To enhance the water content of these lenses, two basic formulations are employed. In one case a hydrophilic monomer such as HEMA is combined with a highly hydrophilic charged (ionic) monomer. As a consequence of the strong attractive interactions between water and the charged functionality, the water content of the lens is greatly increased, increasing the pliability and comfort level of the lens. Alternatively, the lens can also be designed by combining two highly hydrophilic nonionic monomers.

The Chemical Details

Cross-linked polymeric materials with optical transparency and biocompatibility are used to construct hard contact lenses. The monomers commonly used in hard contact lenses possess a high degree of hydrophobicity due to their inability to form hydrogen bonds with water. The ester methyl methacrylate (MMA) (Fig. 14.6.1), $CH_2C(CH_3)COOCH_3$, was the first monomeric unit used in 1948.

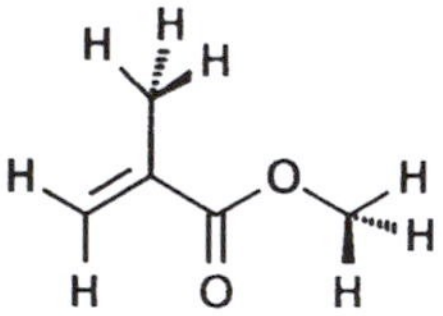

Figure 14.6.1 ▶   The molecular structure of the ester methyl methacrylate (MMA).

Lenses with a greater degree of gas permeability were designed in the mid-1970s using siloxane-based monomers. For example, a copolymer of methyl methacrylate and the monomer known as methacryloxypropyl tris(trimethylsiloxysilane) or TRIS (Fig. 14.6.2) was formulated in 1975 to provide a number of desir-

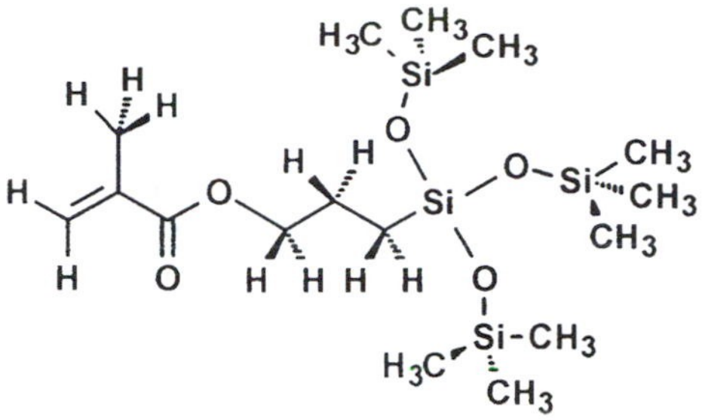

Figure 14.6.2 ▶ The molecular structure of the monomer methacryloxypropyl tris(trimethylsiloxy silane) or TRIS.

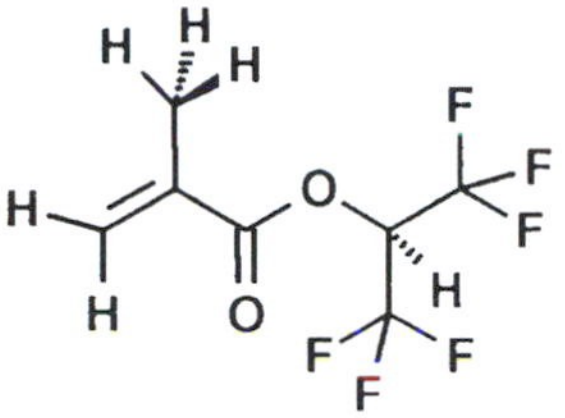

Figure 14.6.3 ▶ The molecular structure of the monomer hexafluoroisopropylmethacrylate (HFIM).

able features such as oxygen permeability, wettability, and scratch resistance.[1] In the 1980s approaches using a number of fluorine-based monomers were successful. A polymer of MMA, TRIS, and hexafluoroisopropylmethacrylate (HFIM) (Fig. 14.6.3) is one such formulation for hard lenses. The hydrogel poly(hydroxyethyl methacrylate) composed of cross-linked monomers of 2-hydroxyethyl methacrylate (HEMA) (Fig. 14.6.4) was the first soft lens material. (*Cross-linking*

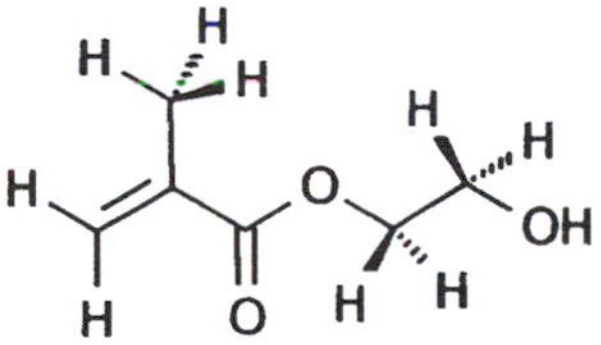

Figure 14.6.4 ▶ The molecular structure of the monomer 2-hydroxyethyl methacrylate (HEMA).

consists of bonding between the main polymer chains to add strength to the material.) The presence of the hydroxy functional group (i.e., –OH) permits hydrogen bond formation with water and leads to the capacity to absorb water (typi-

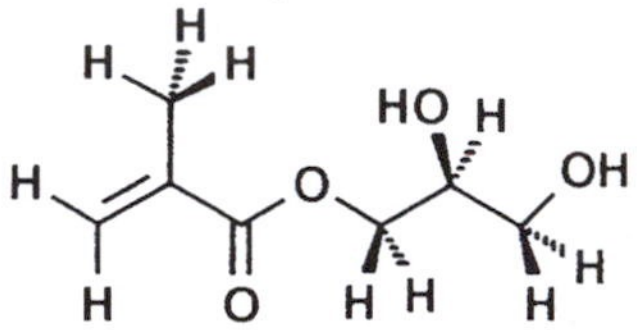

Figure 14.6.5 ▶ The molecular structure of the hydrophilic nonionic monomer 2,3-dihydroxypropylmethacrylate.

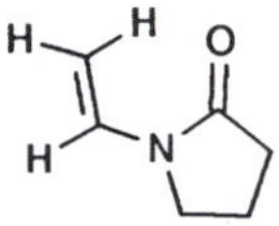

Figure 14.6.6 ▶ The molecular structure of the hydrophilic nonionic monomer N-vinyl-2-pyrrolidinone (NVP).

cally 38% by weight).[2] Highly hydrophilic nonionic monomers used in hydrogel lenses include glycerol methacrylate (GM) or 2,3-dihydroxypropylmethacrylate (Fig. 14.6.5), N-vinyl-2-pyrrolidinone (NVP) (Fig. 14.6.6), and N,N-dimethylacrylamide (DMA) (Fig. 14.6.7). The Accuvue lens currently manufactured by Johnson and Johnson is an example of a hydrogel polymer with an ionizable monomer (hence enhanced water absorption). Cross-linking HEMA (Fig. 14.6.4) and methacrylic acid (MAA) (Fig. 14.6.8) leads to a water content of 58%. Other combinations of polymers (such as methyl methacrylate and polyvinyl pyrolidone) lead to soft hydrophilic lens materials with water contents as high as 70%.[3,4]

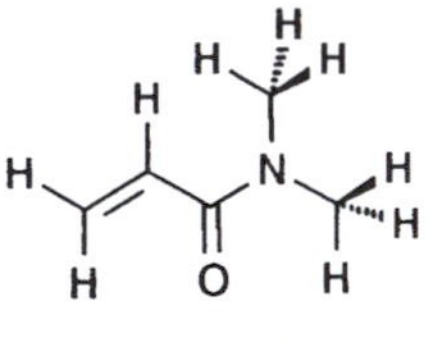

Figure 14.6.7 ▶ The molecular structure of the hydrophilic nonionic monomer N,N-dimethylacrylamide (DMA).

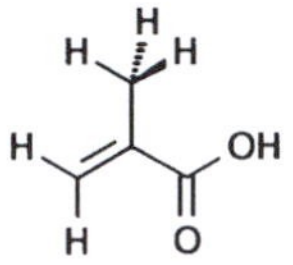

Figure 14.6.8 ▶ The molecular structure of the monomer methacrylic acid (MAA).

KEY TERMS:	hydrophobic	hydrophilic	monomer	cross-linking
	polymer	copolymer	hydrogel	

References

[1] "Contact Lens Material." Dr. Jay F. Kunzler and Dr. Joseph A. McGee, Department of Polymer Chemistry, Bausch & Lomb, *Chemistry & Industry* **21** (1995), 615.

[2] "History of Contact Lens." Indiana University School of Optometry, V232 Contact Lens Methods and Procedures, Fall 2001,
http://www.opt.indiana.edu/v232/lectures/history/History.ppt

[3] Vision Care, ICN Pharmaceuticals, Inc., http://www.icnvisioncare.com/allabout.html

[4] Generic Soft Lens Materials, Visiontech Services, http://www.vstk.com/softLens.htm

Related Web Sites

► "Contact Lenses Identified by Type of Material." CIBA Vision,
http://www.cibavision.com/text/forsight/allabout/C04.03.05.html

Other Questions to Consider

7.6 How is a fabric made water-repellent or waterproof? *See* p. 89.

7.7 How does a bullet-proof vest work? How is it made? *See* p. 92.

7.8 Why is cotton so absorbent and why does it dry so slowly? *See* p. 95.

9.5 Why are floor waxes removable with ammonia cleansers? *See* p. 124.

12.6 What is the purpose of the thread that runs vertically through the clear field on the face side of U.S. currency? *See* p. 152.

13.1 How do sutures dissolve? *See* p. 166.

Chapter 15 ▶

Connections to Materials

15.1 What Is the Composition of an Artificial Hip?

The success of hip replacements has been greatly advanced by major development in the biomaterials for orthopedic devices. What aspects of chemistry must be considered in designing an effective artificial hip?

The Chemical Basics

Biomaterials are synthetic and naturally occurring materials that are foreign to the body but are used to replace a diseased organ or tissue or augment or assist a partially functioning organ or tissue. Cardiovascular, orthopedic, and dental applications are some of the most common areas in which biomaterials are employed.

Successful applications of materials in medicine have been experienced in the area of joint replacements, particularly artificial hips. As a joint replacement, an artificial hip must provide structural support as well as smooth functioning. Furthermore, the biomaterial used for such an orthopedic application must be inert, have long-term mechanical and biostability, exhibit biocompatibility with nearby tissue, and have comparable mechanical strength to the attached bone to minimize stress. Modern artificial hips are complex devices to ensure these features.

Designed as a cup and ball joint, the artificial hip or prosthesis consists of two parts: the stem or femoral component and the socket or acetabular component. Inserted into the thigh bone or femur, the metal stem is composed of a titanium alloy or a cobalt–chromium–molybdenum alloy with a metal or ceramic ball-shaped head. Anchored to the hip bone, the acetabular component is a metal hemispherical shell with a plastic lining to act as a bearing for the femoral head. Figure 15.1.1 illustrates the two structural components of the artificial hip joint. Both the stem and the socket may be cemented into place with a special epoxy-type cement to bond the metal component to the bone. Alternatively, newer cementless surgical procedures use a porous coating of a calcium-phosphate-based ceramic to promote bone growth into porous surfaces. Hip implants of the future

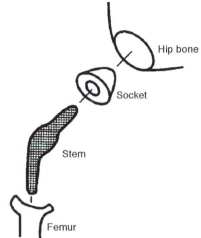

Hip bone

Socket

Stem

Femur

Figure 15.1.1 ► A schematic diagram of the two components of an artificial hip: the stem or femoral component and the socket or acetabular component.

will replace the metal components with fiber composites that will more evenly match the stiffness of bone.

The Chemical Details

The composition of bone is primarily a two-phase system of collagen, an organic substance, and calcium hydroxyapatite [$Ca_{10}(PO_4)_6(OH)_2$], an inorganic calcium phosphate mineral. Variations in the mineral–collagen ratio lead to changes in physical properties. Increased proportions of collagen tend ultimately to greater flexibility, while a higher level of mineral leads to increased brittleness. At the present time, no synthetic material mimics perfectly the mechanical properties of bone. Metals such as titanium alloys or cobalt–chromium–molybdenum alloys have the necessary strength and fracture resistance but have high degrees of stiffness. Ceramics such as hydroxyapatite generally possess the hardness and compressibility of bone but not the fracture resistance. Polymeric materials such as polymethylmethacylate and polyethylene have improved performance due to low stiffness and reasonable fracture resistance but generally lack in strength. In addition to mechanical properties, the biocompatibility and chemical degradation properties of biomaterials for implants must also be addressed. For example, the selection of metals such as titanium ensures a high degree of corrosion resistance in physiological media (i.e., water with dissolved oxygen, hydronium ions, chloride ions, etc.). Titanium derives its corrosion resistance by forming a protective oxide surface coating of TiO_2. While a material with properties identical to bone has not yet been devised, the example of the design of biomaterials for total hip replacement illustrates the chemical and material advances that ultimately improve the quality of human life.

KEY TERMS: biomaterial   alloy

Related Web Sites

▶ "New wedge-shaped hip stem to enter market." Orthopedics Today, 1996, http://www.slackinc.com/bone/ortoday/ot96toc.asp

▶ "Case Study of Materials Selection for Total Hip Replacement." Materials Department, Queen Mary and Westfield College, University of London, http://www.materials.qmul.ac.uk/casestud/implants/

▶ "A Patient's Guide to Artificial Hip Replacement." Randale Sechrest, MD, Medical Multimedia Group. http://www.medicalmultimediagroup.com/pated/joints/hip/ hip_replacement.html

▶ "Metal-on-Metal: A Potential Alternative to Polyethylene." *Orthopedics Today*, http://www.slackinc.com/bone/ortoday/ot96toc.asp

▶ "Hydroxyapatite Crystals / Prosthetic Coating." *Wheeless' Textbook of Orthopaedics*, C. R. Wheeless, M.D., 1996. http://www.medmedia.com/o2/49.htm

▶ "Titanium." *Wheeless' Textbook of Orthopaedics*, C. R. Wheeless, M.D., 1996, http://www.medmedia.com/o16/69.htm

---

15.2   What Is Liquidmetal (in Liquidmetal Golf Clubs)?

"Breakthrough technology that will change the game forever."[1] This is the claim of the manufacturers of an innovative combination of metals revolutionizing the golf industry. Inserted into the clubheads of putters, irons, and drivers, this material provides golfers with more solid shots with greater energy upon impact and enhanced accuracy upon trajectory than conventional clubs. What is the chemistry of this unique material?

The Chemical Basics

Liquidmetal is a patented five-metal alloy made by combining nickel, zirconium, titanium, copper, and beryllium. Liquidmetal is the discovery of two scientists from the California Institute of Technology in Pasadena. Dr. Atakan Peker, Senior Scientist, and Dr. William L. Johnson, Mettler Professor of Materials Science at Caltech, invented Liquidmetal in the spring of 1992. Liquidmetal Golf, the Laguna Niguel, CA company that markets golf clubs made with the alloy, claims that the structure of this innovative combination of metals produces a stronger, lighter, more resilient material that transfers more energy to a golf ball upon impact.

The Chemical Details

Liquidmetal is a bulk amorphous or noncrystalline alloy that contains approximately two-thirds zirconium as well as nickel, titanium, copper, and beryllium. In the materials world, this type of alloy is often described as a liquid metal, a metallic glass, or a "glassy metal." The origin of this terminology lies in the microstructure that the alloy adopts as it is cooled from a molten state. Conventional metals and alloys form crystalline structures with defined three-dimensional arrangements of atoms or molecules. By cooling the molten form of Liquidmetal rapidly, the metallic atoms in Liquidmetal do not arrange in an ordered manner but in random fashion. This *amorphous* state is known as a *glassy* state, as glassy substances lack the rigid regularity of a crystalline solid. As the molecules of liquids also are not locked into definitive locations, the alloy thus has properties characteristic of liquids, hence the name Liquidmetal.

Why does this amorphous structure improve the performance of Liquidmetal? Crystalline materials are actually composed of many small crystals held together in a polycrystalline structure. Each single crystal containing a regular arrangement of atoms is known as a grain. While grains are packed together tightly, the shapes of grains are irregular and lead to *grain boundaries* when the crystals come into contact. The grain boundaries of conventional crystalline metals lead to microscopic gaps and pockets that are potential sites of weakness. The amorphous structure of Liquidmetal means that there are no grains and no grain boundaries to cause defects in the alloy's structure.

The composition of an alloy dictates its particular properties. Liquidmetal has an extremely high tensile strength of 1900 MN m^{-2}, twice that of titanium or stainless steel.[1,2] The manufacturer also claims a high degree of resiliency for Liquidmetal, two to three times higher in resistance to deformation than conventional metals.[1] Combined with its low density[2] of 6.1 g cm^{-3} (lower than stainless steel but higher than titanium[1]), the alloy has a very high strength-to-weight ratio compared to other materials. The application in golf club heads takes full advantage of this strength-to-weight ratio. The claim that more energy is transferred to the ball upon impact results from the metal absorbing less energy than conventional materials on impact. More power leads to greater distances per shot. The alloy also possesses exceptional vibrational damping qualities, reducing the shock on impact and providing a soft but solid feel. The alloy also exhibits a high degree of hardness, an important feature to reduce wear.

KEY TERMS: alloy   amorphous   crystalline   glass

References

[1] Properties of Liquidmetal Alloy, Liquidmetal Golf,
     http://www.liquidmetal.com/index.cfm?action=tech

[2] Liquidmetal Swings into Action, *Materials World* **6** (1998), 547–548.

15.3 | How Does an Automatic Fire Sprinkler Head Operate?

Many office buildings rely on ceiling sprinkler systems as effective means of extinguishing a fire. How are these systems activated? The heat generated by a fire is the key to the operation of such water sprinklers.

The Chemical Basics

Automatic fire sprinklers are inserted into a network of piping that contains water under pressure. The sprinklers operate individually by heat-activation. As a fire's intensity increases, the heat of a fire builds, exposing the sprinkler head to elevated temperatures generally between 135 and 225°F. A typical operating temperature for a sprinkler head is 165°F,[1] although temperature ratings as low as 135°F and as high as 360°F are available.[2] When the operating temperature is reached, either a solder link melts or a liquid-filled glass bulb shatters. In the solder link operating mechanism, two pieces of metal are held together with a special solder, creating the "link." The link is connected to a mechanical mechanism that holds a cap or plug in place in the water pipe. As solder melts at a specific temperature, the cap over the orifice or nozzle is ejected by water pressure. When the sprinkler head is opened, water is directed through a deflector to create a spray that is released directly over the source of heat. In glass bulb sprinklers, liquid within a fragile glass bulb expands in volume as its temperature is raised by the heat of the fire. The glass bulb also holds a cap or plug in place in the water pipe. At a particular temperature, the volume has increased enough to shatter the glass bulb, triggering the discharge of water. Also used are chemical pellets that melt at a predetermined temperature. Fusible alloys are not only used in fire sprinkler heads but also in thermal alarms and in fuses interrupting an electrical circuit when the current becomes excessive.

The Chemical Details

Alloys are mixtures of metals combined to obtain specific characteristics and enhanced properties for a particular application. The term *fusible metals* or *fusible alloys* denotes a group of alloys that have melting points below that of tin (232°C, 449°F). Most of these substances are mixtures of metals that by themselves have relatively low melting points, such as tin, bismuth (m.p. 275°C), indium (157°C),

TABLE 15.1 ▶ Melting Points of Fusible Alloys of Variable Composition

Bi	Cd	In	Pb	Sn	Melting point (°C)	Melting point (°F)
45	5	19	23	8	47	117
49		21	18	12	58	136
50	10		27	13	70	158
42	9		38	11	70–88	158–190
52			32	16	95	203
55			45		124	255
58				42	138	280
40				60	138–170	280–338

cadmium (321°C), and lead (327°C). Table 15.1 below illustrates the dependence of melting point on the composition of several fusible alloys.[3]

Note that all combinations of metals yield an alloy with a melting point significantly lower than that of any of the individual metals. Even more extensive listings of fusible alloy compositions and melting points are available.[4] As a point of reference, paper will spontaneously begin to burn when heated to 451°F, just above the melting point of tin.

KEY TERMS: alloy

References

[1] Reliable Model ZX-QR-INST Pendent Horizontal Sidewall Quick Response Institutional Sprinkler, Reliable Automatic Sprinkler Co., Inc., http://www.reliablesprinkler.com/sprinklr/137c.htm

[2] "Reliable Model F1 and Model F1FR Intermediate Level Sprinklers." Reliable Automatic Sprinkler Co., Inc., http://www.reliablesprinkler.com/sprinklr/170b.htm

[3] Fusible Alloys, Clad Metal Industries, http://www.cladmetalindustries.com/Alloys.html

[4] Fusible Alloys with Characteristics, Canfield Technologies, http://www.solders.com/aloyprod.htm

Related Web Sites

▶ "An Introduction to Automatic Fire Sprinklers, Part 1." Nick Artim, *WAAC Newsletter* **16** (3) (Sept. 1994), 20, http://sul-server-2.stanford.edu/waac/wn/wn16/ wn16-3/wn16-309.html

▶ "An Introduction to Automatic Fire Sprinklers, Part II." Nick Artim, *WAAC Newsletter* **17** (2) (May 1995),
http://palimpsest.stanford.edu/waac/wn/wn17/wn17-2/ wn17-206.html

▶ "Star Galaxy? SGQR Quick Response Conventional Sprinkler." Star Sprinkler, Inc., http://www.starsprinkler.com/Tech_Data/SPECS/11411.PDF

Other Questions to Consider

| 5.2 | Why do homemade copper cleaners use vinegar? *See* p. 36.

Chapter 16 ▶

Connections to Biochemistry

Previous Questions to Consider

| 5.9 | Why are bombardier beetles known as "fire-breathing dragons"? *See* p. 49. |

| 5.15 | What is the source of Dr. Seuss' green eggs (and ham)? *See* p. 59. |

| 7.10 | How does milk froth for cappuccino coffee? *See* p. 100. |

| 7.11 | Why is sickle cell anemia a "molecular disease"? *See* p. 102. |

| 13.3 | How do forensic chemists use visible stains to trap thieves? *See* p. 170. |

| 13.7 | Why should you avoid acidic foods when taking penicillin? *See* p. 178. |

| 13.15 | What causes ice cream to develop a gritty texture during long periods of storage? *See* p. 197. |

| 13.16 | Why is some olive oil designated as "extra virgin"? *See* p. 199. |

| 13.17 | How can sucralose, an artificial sweetener made from sugar, contain *no* calories? *See* p. 205. |

Chapter 17 ▶

Connections to the Environment

Previous Questions to Consider

| 5.8 | How is lime used to mitigate the acid rain problem? *See* p. 46.

| 5.10 | Why do seashells vary in color? *See* p. 51.

| 8.1 | What do meteorologists use to "seed" clouds? *See* p. 105.

| 8.2 | Why do citrus growers spray their trees with water to protect them from a freeze? *See* p. 107.

| 9.4 | Why do hydrangeas vary in color when grown in dry versus wet regions? *See* p. 122.

| 12.3 | Why is the aurora borealis so colorful? *See* p. 145.

| 13.8 | Why do we detect odor in an "odorless gas" leak? *See* p. 180.

| 13.14 | Are flamingos naturally pink? *See* p. 196.

Index

Absorber material, 157
Absorption
    and hydrogen bonding, 95
Absorption spectrum
    vs emission spectrum, 159
Acetaminophen, 116
    chemical structure, 117
Acetic acid, 37, 46, 53, 60
    equilibrium constant of, 73
Acetylsalicylic acid, 116
    chemical structure, 117, 173
Acid rain, 38, 46–48
Acid/base content, 72, *see also* pH
    indicators of, 74, 75
Acrylics, 125
adiabatic, 134
Adsorption
    in dessicants, 31
    vs absorption of chemical pollutants,
        47
    water, to silica microspheres, 27
Albumen, 22
Alcohols, *see also* Ethanol
    polyfunctional, 24
    polyhydric, in cold cream, 9
Alkali metals
    thermal ionization, 3
Alkalinity, 72, *see also* pH
    of drain cleaners, 126
Alkyl sulfates, 98
Alloys
    fusible, 230, 231
    iron and tinplate, 43
    liquidmetal, 228, 229
    oxidation of, 37, 43
Alpha hydroxy acids (AHAs)
    chemical structure, 184
    in antiwrinkle creams, 183

Alpha hydroxycarboxylic acid, 55
Alpha-hydroxyethanoic acid
    chemical structure, 184
Alumina, 35, 36
Aluminosilicates
    glass, 76
    zeolites, 31
Aluminum
    and alumina, 35
    concentration and hydrangea color,
        122
    in flashbulbs, 42
    oxidation in fuel, 36
    oxides, 42
    silicates, 57
Amidotrizoates, 129
α-Amino groups, 170
Amino substituents, 165
7-Amino-4-methylcoumarin, 159
    absorption/emission spectra, 160
    chemical structure, 160
Aminophenol, 188
    chemical structure, 189
Ammonia, 60
    and bleach, 182
    chloramine species, 182
    cleaners, 124, 182
    gas and baking, 67
    hydrazine, 182
Ammonium lauryl sulfates, 98, 214
    chemical structure, 98, 215
Ammonium perchlorate, 35
Amorphous structure, 26
Anesthetics, 78, 79, 81
Anionic surfactants, 98
Anode, 44
Anthocyanins, 123
    chemical structure, 123

chemical structure, 124
Anticaking agents, 29
Antistatic compounds, 86
Aramid fibers, 92
Aromatic ring structure, 98, 163
Arrhenius law, 137
Arrhenius, Svante, 72
Artificial hip
structure of, 226–228
Ascorbic acid, 37, 184
chemical structure, 185
Aspirin, *see* Acetylsalicylic acid
chemical structure, 173
website reference, 175
Astaxanthin
chemical structure, 196
Aurora borealis/australis, 145, 146

Bakers' yeast, 68
Baking soda, *see* Sodium bicarbonate
Barium
in contrast media, 2
Barium sulfate, 2
in contrast media, 17
solubility of, 18
Basic lead sulfate, 57
Benzene, 163
absorption spectrum, 163
ring structure of, 93
solvent for illegal drugs, 178
Benzophenones, 162
Beta hydroxy acids (BHAs), 183, 184
chemical structure, 184
Biomaterials, 166
artificial hips, 226
Biphenyl, 159
chemical structure, 160
9,10-Bis(phenylethynyl)anthracene, 141
chemical structure, 142
Bismuth
melting point of, 230
Bisphenol A
chemical structure, 217
polycarbonate, 217
synthesis, 216–218
Bleach, 40
and ammonia, 182
and counterfeit banknotes, 153

household, 182
vs optical brighteners, 158
Bone, 226
absorption of X-rays, 1
and artificial hips, 226
composition of, 227
Borax, 55
Borosilicate glass, 76
Brass
surface corrosion, 37
Bromine
in halogen lights, 110
Bubbles
in carbonated beverages, 11
spherical shape of, 84, 85
in colloidal dispersion (foam), 21
Bunsen, Robert, 42
Butane, 180
1,4-Butanediol, 188
chemical structure, 189
2-Butene-1-thiol, 169
chemical structure, 169
tert-Butyl mercaptan
chemical structure, 181
Butylene glycol, 188
chemical structure, 189

Cadmium
melting point of, 231
paint pigments, 58
Calcium carbonate
action of vinegar on, 53
and soap scum, 54
chalk, 53
composition of eggshells, 53
composition of pearls, 26
composition of sea shells, 51
limelight and, 66
limestone, 48, 61
precipitation reaction, 51
solubility product of, 51, 61
Calcium chloride (calcinated lime)
deliquescent nature of, 31
Calcium hydroxyapatite, 227
Calcium hypochlorite, 70
Calcium oxide (lime)
and acid rain, 47
as a desiccant, 31

crystallization of, 65
incandescence of, 65
website reference, 49
Calcium silicate, 29, 30
Calcium sulfate
as desiccant, 31
gypsum, as water softener, 61
gypsum, weathering of, 61
hemihydrate of, 31
Canthaxanthin, 196
chemical structure, 197
Capillary condensation, 31
Capric acid, 54
Caprylic acid, 54
Capsaicin, 83
chemical structure, 84
Capsaicinoids, 83
Carbon dioxide, 13
and acid rain, 48
and ammonia, 67
and baking soda, 67
and carbonate ions, 52
and carbonation, 84
and carbonic acid, 116
dissociation in seawater, 52
and carbonic ion, 61, 75
and disappearing ink, 74
and natural gas, 180
antacids and, 116
as a fire extinguisher, 39, 63
atmospheric, undissolved, 53
carbonated beverages, 11
corrosion and, 37
degradation of PGA, 167
fermentation and, 176
free-radical polymerization, 215
in cloud seeding, 106
in fermentation, 68
porosity of eggshells to, 5
shape of bubbles, 85
solubility in water, 12
illustration, 12
synthesis of bisphenyl A, 217
Carbonate white lead, 57
Carbonation
and bubble shape, 84
and warming, 11
process, 11–13

Carbonic acid
action on limestone, 61
antacids and, 116
carbonated beverages and, 13
decomposition in stomach, 117
dissociation of, 51
equilibrium constant of, 117
esters of, 16
ink and, 74
ionization and hydronium ion, 75
linear chains of polycarbonates, 217
Carboxyl groups, 93, 121
absorption wavelengh of, 165
deprotonation of, 210
l-Carvone
chemical structure, 193
l-Carvone, 192
CAT scans, *see* Computerized axial tomography
Cation exchange
in water softeners, 54
Cationic surfactants, 98
Cellulose, 46
absorption wavelength, 158
hydrogen bonding in, 96, 97
hydroxyl group in, 90
repeating monomeric units, 209
structure of, 90, 96
Cetyl alcohol (hexadecanol), 187
chemical structure, 187
CFC-113, 88
chemical structure, 88
Chain-growth polymerization reaction, 214
Chain-transfer agent, 214
Charles' Law, 6
Charles, Jacques, 6
Chelation, 121
Chelators, 54
Chemiluminescence, 139
Chemoreception, 190
Chloramines, 182
Chlorine, 16, 36
in swimming pools, 70, 71
Chloroallyl methanamine chloride (Quaternium-15), 98
chemical structure, 99
Chlorofluorocarbons, 88
Chloroform, 8, 78

chemical structure, 79
Chromatophores, 52
Chrome, 37
 pigments, 57
Chromic acid
 as oxidizing agent, 46
Chromophore, 163
Cineole, 191
 chemical structure, 192
Cinnamates, 162
Cinoxate, 162
 chemical structure, 164
Citric acid, 37, 116, 117, 121, 183
 chemical structure, 117, 122, 185
Citrus crops
 freezing weather and, 107, 108
Clausius–Clapeyron equation, 68
Cloud condensation nuclei, 105
Cloud seeding, 105, 106
Coacervation (phase separation), 212, 213
Coalescence, 105
Cocaine
 chemical structure, 177
Coefficient of thermal expansion, 112
Cold stabilization, of wine, 14
Collagen, 227
Colligative property, 20
Colloidal dispersion, 22
Combustion
 decomposition of sodium carbon-
  ate, 48
 firefighting, 63
 metals, 39, 42
 zirconium, 42
common ion effect
 solubility of, 18
Complex formation, in cyanide poison-
  ing, 119
Computerized axial tomography (CAT),
  1, 2, 128, 129
Conalbumin, 22
Condensation, 68
 capillary, in desiccant, 31
 cloud nuclei, 105
 raindrop formation, 105
Condensation reaction
 in ether synthesis, 173
 in Kevlar synthesis, 93, 94

polymers, 96, 209
Conjugate base, 74
Conjugation of double bonds, 163
Contrast media, 1, 2
Coordinate covalent bonds, 119
Coordination complex, 119
Coordination number, 65
Copolymers, 125
 in contact lenses, 222
 in sutures, 167
Copper
 alloys in organ pipes, 113
 chelates, 55
 cleaners of, 36–38
 ions in food, 120
 oxidation products, 37
 pigments, 57
 thermal conductivity, 144
 usage in mirrors, 143
Cormack, Allen, 1
Correction fluids, 8
Cotton
 absorbency of, 95–97
 composition of, 46
 dyeing of, 46
 fibers, hydrogen bonding of, 90
 filaments in lightbulb, 110
 in Scotchgard, 89
 rag paper as currency fabric, 149,
  150
Coumarin, 159
 chemical structure, 159
Crane & Company Inc., 149
Cream of tartar (potassium bitartrate), 14,
  21, 22, 67
Cristobalite, 26
Cross-linking polymer, 125
Crystals/crystallization
 calcium oxide, 31, 65
 diamond, 32
 glass, 75, 76
 gray tin, 113, 114
 ice, 20, 106
 mica, 147
 thermoplastics, 92
 titanium dioxide in pigments, 150
 white tin, 113, 114
 wurtzite structure, 106

Currency
    cotton/linen, 149, 150
    optically variable ink in, 152, 153
    security thread in, 152
Cuttlefish (*Sepia officinalis*), 52
Cyanide
    pigments, 57
Cyanide poisoning, 118–120
    antidote, 119
    mechanism, 119
Cyanoacrylates, 219
    chemical structure, 220
Cysteine, 60
    chemical structure, 60
Cytochrome oxidase
    role in cyanide poisoning, 118

α-D-glucose
    chemical structure, 198
α-D-glucose
    chemical structure, 198
β-D-glucose
    chemical structure, 198
Dacron^{TM}
    linked monomers
      structure of, 97
    repeating unit
      structure of, 96
Dacron^{TM}
    repeating unit, 96
Decomposition reactions, 41
    ethers, 173
    fermentation, 68
    hydrolysis of sucralose, 206
    thermal, 63, 65
Deliquescence
    of calcium chloride, 31
Density
    fog effect, 24
    glass properties, 76
    lava lamp mode of action, 15
    of cross-links in polymers, 125
    of ink flakes and optical effect, 156
    of liquidmetal, 229
    tissues and X-rays, 1, 128
    water, 136
Desflurane, 79
    chemical structure, 81

Desiccants, 30, 31
Di-(4-tert-butylcyclohexyl)peroxydicarbonate,
    214
    chemical structure, 215
Diamines, 93
Diamond
    color of, 32
Diatrizoate anion, 129
    chemical structure, 129
Dibasic acids, 153
Dibenzoylmethanes, 162, 163
Dicarboxylic acid, 93
Dichloramine, 182
Dicobalt edetate, 119
Dielectric material, 157
16,17-Dihexyloxyviolanthrone, 141
    chemical structure, 142
Dihydrocapsaicin, 83
    chemical structure, 84
2,3-Dihydroxypropylmethacrylate, 224
    chemical structure, 224
N,N-dimethylacrylamide (DMA)
    chemical structure, 225
N,N-dimethylacrylamide (DMA), 224
Dioxybenzone, 162
    chemical structure, 162
9,10-Diphenylanthracene, 141
    chemical structure, 142
Disappearing ink, 73–75
Dispersion
    colloidal, 22
    of water in fog, 24
Dispersion forces, 86
Disproportionation, 41
DMA, *see N,N*-dimethylacrylamide
Dodecanethiol, 214
    chemical structure, 215
Double bonds, conjugation, 159, 163, 201
Drain cleaners, 126, 127
Dust particles, 86

Edison, Thomas, 109, 110
    website reference, 77, 111
EDTA (ethylenediaminetetraacetic acid),
    119
    chemical structure, *see* color fig-
      ures, 121
    formation constants, 119

in food products, 120
Electrical charge, 86
Electron microscopy, 27
Emission spectrum, vs absorption spectrum, 159
Emission spectrum, vs. absorption spectrum, 3
Emulsions
skin care, 9
skin care, website reference, 190
solid, 26
website reference, 10
Endothermic reactions, 9, 64, 135
Enflurane, 79
chemical structure, 80
Enteric coatings, 209
Enthalpy, 64
of fusion, 108
of vaporization, 68
Entropy
of vaporization, 68
ePTFE (expanded polytetrafluoroethylene),
91
Equilibrium constant
dissociation of acetic acid, 73
in formation of carbonic acid, 117
in solubility of magnesium hydroxide, 19
Essential oils, 190
Ester
chemical structure (general), 172
Ester linkages, polyester-forming, 153
Ethane, 180
Ethanethiol, 181
chemical structure, 181
Ethanol
absence of, in cosmetics, 186
chemical structure, 189
flash point of, 7
freezing and boiling points of, 112
freezing point of, 112
in baking, 68
in cold creams, 9, 10
in disappearing ink, 74
in fermentation, 176
in thermometers, 112
in wine, 13, 14
solvency of, 188

vapor pressure of, 9
Ether (diethyl ether), 78
chemical structure, 79
Ethyl chloride, 79
chemical structure, 80
Ethylenediaminetetraacetic acid (EDTA),
119
chemical structure, 121
formation constants, 119
in food products, 120
16,17-(1,2-Ethylenedioxy)violanthrone, 141
chemical structure, 142
Ethylhexyl salicylate, 163
chemical structure, 164
Evaporation
in batteries, 137
in thermal fog machines, 23
of alcohols, 9
of ammonia, 125
of water in clouds, 105
Exfoliators, 183
Exothermic processes, 39, 49, 108, 133
Expanded polytetrafluoroethylene (ePTFE),
91

Fabric
bullet-resistant, 92, 93
website reference, 94
cotton, rate of drying, 95, 97
cotton/linen, of currency, 149, 150
dyes in cotton, 46
water-repellent/waterproof, 89–91
website reference, 91
Face-centered cubic (fcc) lattices, 33, 65
Fahrenheit, Daniel Gabriel, 112
Fatty acids
carboxylic acids, 201
chemical structure, 202
composition of plant oils, 203
general chemical structure, 201
hydrogen peroxide formation, 50
monoglycerides and diglycerides,
101
soap production, 54, 55
Fermentation
in baking, 68
secondary, in vinegar, 175
wine production, 13, 14

Ferric hydroxide, 57
Fibers, natural, yellow tinge of, 158
Fire
 class D, 39
 extinguished by baking soda, 63, 64
 flash point, 7
 magnesium, 39, 40
Fire sprinkler heads, 230, 231
 website references, 231
Flash point, 7
Flashbulbs, photographic, 39, 42
Fluorescence
 dyes, 140, 141
 in currency fibers, 153
 inks, 153
 optical brighteners, 158
Fluorocarbons
 in water-resistant/waterproof fabrics, 88
Fluorosilicic acid, 76
 chemical structure, 76
Foams, 22
Fog
 machine made, 23, 24
 vs. smoke, 23
 website reference, 25
Formation constant, EDTA, 119
Free-radical polymerization reaction, 214
Freezing point depression constant, 20
Freezing process, 20
 enthalpy change, 108
Fry, Art, 214
Furfurylthiol, 168
Fuwa, Kyozo, 75

Galen, 166
Galileo, 111
Gas permeability, of contact lenses, 221
Gay-Lussac, Joseph, 6
Glass
 frosting with hydrofluoric acid, 75
 shatterproof, 216, 217
 thermal expansion of, 112
Glassy metals, 228, 229
Glutamic acid, 103
 chemical structure, 103
Glycerin, 25, 188

 chemical structure, 189
Glycerol
 chemical structure, 201
Glycerol aminobenzoate, 162
 chemical structure, 163
Glycerol methacrylate (GM), 224
Glycolic acid, 54, 166, 184
 chemical structure, 167
 copolymers of, 167
 in sugar cane, 183
 website reference, 55
Glycols, 153
Glycosidic linkage, 199
Gold films, 144
Groundwater, 52–54, 61

Hailstones, 106
Hale-Bopp comet, 3
Halogen lights, 110
 website reference, 111
Halogens
 binary acids of, 76
 hydrocarbons and, 88
Halothane, 79
 chemical structure, 80
Hard water, 52–54
Heat capacity, 108, 133, 136
HEMA, *see* 2-Hydroxyethyl methacrylate
Henry's Law, 12
Heroin
 chemical structure, 177
Heterochain polymers, 217
Heterogeneous reaction, 64
Hexadecanol, *see* Cetyl alcohol
Hexafluoroisopropylmethacrylate (HFIM), 223
 chemical structure, 223
1,6-Hexamethlenediamine, 93
HFIM, *see* Hexafluoroisopropylmethacrylate
Homocapsaicin, 84
Homodihydrocapsaicin, 84
Homosalicylate, 163
 chemical structure, 164
Hope diamond, 32
Hounsfield, Godfrey, 1
Hydrangeas, 122, 123

Hydrazine, 182
Hydrocarbons, 180, 201
    as fire component, 63
    dispersion forces of, 86
    halogenated, 79
    in natural gas, 180
Hydrofluoric acid, 75
Hydrogen bonding, 15, 41, 83, 85
    in Kevlar synthesis, 93
    in polymer–surface bonding, 220
    loss of, in water-resistant fabrics,
        90
    of cellulose, components of, 95, 97
    of cotton fibers, 46
    of hydrophilic substances, 88
    of polyfunctional alcohols, 24
Hydrogen peroxide, 40, 41
    chemical structure, 40
Hydrogen sulfide, 37, 57, 59, 60, 168,
        180
Hydrolysis, 166, 167, 172
    of a triglyceride, 201
    of amide linkages, 179
    of an ester, 177, 201
Hydrophobic force, 15
Hydrophobicity
    of hard contact lenses, 222
    of surfactants, 97
N-[4-Hydroxy-3-methoxyphenyl)methyl]-
        8-methyl-6-noneamide, see Cap-
        saicin
Hydroxyaniline, see Aminophenol
    chemical structure, 189
2-Hydroxyethyl methacrylate (HEMA),
        222
Hydroxyl groups, 90, 123
    acetylation, 177
    in alcohols, 173, 217
    in alpha hydroxy acids, 183
    in benzophenones, 163
    in cellulose-based fibers, 90
    in glycols, 153
    in leucoindigo, 46
    in sugars, 22
    removal of, 196
Hydroxypropyl methylcellulose phthalate
        (HPMCP), 209
Hygroscopicity, 22, 24, 31

    of propylene glycol and triethylene
        glycol, 24

Ideal gas law, 69
Imidazole, 159
    chemical structure, 159
Incandescence, 65, 110
Indigo, 45
    chemical structure, 45
Indigo dye, 45
    website reference, 46
Indium, melting point, 230
Infrared radiation, 110, 130, 144
Ink
    disappearing, 73–75
    sepia, 52
Intermolecular forces, 15, 68, 79, 83, 85,
        95, 204
    dispersion, 86
    in denatured proteins, 101
Iodine
    in halogen lights, 110
    in radiopaque media, 2, 129
Ions/ionization
    aluminate, 127
    atmospheric gases, 145, 146
    carbonate, in formation of calcium
        carbonate, 51–53
    hydrogen, and pH scale, 19, 70, 72,
        73
    metals
        in disproportionation process, 41
        in soap scum formation, 53, 54
    thermal, alkali metals, 3
Iothalamates, 129
    chemical structure, 130
Iridescence, 25–27
Iron oxide, 26, 43, 147
    as catalyst, 35
    in sunblocks, 161, 162
Iron, standard reduction potential, 44
Isoflurane, 79
    chemical structure, 81
Isooctyl acrylate, 214
    chemical structure, 214
4-Isopropyldibenzoylmethane, 163
    chemical structure, 165

Ketamine hydrochloride, 78
  chemical structure, 80
Ketones, 191
Kevlar®, repeating unit and hydrogen
    bonding, 92

Lactic acid, 167, 183
  chemical structure, 167, 168, 185
α-Lactose
  chemical structure, 199
β-Lactose
  chemical structure, 199
Langmuir, Irving, 106
Lapis lazuli, 57
Lattices
  face-centered cubic, 33, 65
    chemical structure, 33, 66
    octahedral holes, 65, 66
    of diamonds, 33
  in zeolite structure, 31
  ionic, of sodium bicarbonate, 64
  similarity of silver iodide to ice, 106
Lauramide DEA, *see* Lauric acid diethanolamine
Lauramidopropyl betaine, 99
  chemical structure, 99
Lauric acid, 54
  sodium salt of, 54
Lauric acid diethanolamine, 188
  chemical structure, 189
Lava lamp, 14–16
Le Chatelier's principle, 18
Lead
  in pipes, 113
  melting point of, 231
  pigments of, 57
Leavening agents, 67
Leucoindigo, 46
  chemical structure, 46
  hydroxyl groups and solubility of,
    46
Light
  absorption vs emission spectra, 159
  and disproportionation process, 41
  color diffraction phenomenon, 156
    illustration of, 156
  from sodium ionization, 3
  interference and iridescence, 26
  polarization of, 192

refraction
  by solutes in ice, 20
  in mother-of-pearl, 51
  in paper currency, 150
  interference and, 148
  refraction/reflection, in pearlescent
    pigments, 147, 148
  ultraviolet, classifications of, 161–
    163
  wavelengths in auroras, 145, 146
Light bulbs
  incandescent, 109
Lightbulbs, *see also* Flashbulbs
  frosted, 75–77
  incandescent, 110
    vs halogen bulbs, 110
Lightstick
  chemical reaction in, 139–141
  uses of, 139
  website references, 143
Lime, calcinated, deliquescent nature of,
  31
Limelights, 64–66
Limestone
  as a sorbent, 48
  composition of, 61
  groundwater and, 52, 53
  hard water and, 61
  thermal decomposition of, 48, 65
Limonene, 87
  chemical structure, 87
l-Limonene
  chemical structure, 193
l-Limonene, 192
Linoleic acid, 54, 203
Liquid-to-gas phase transition, 69
Liquidmetal, 228, 229
Logarithm
  base 10, 72
  natural (base e), 69

MAA, *see* Methacrylic acid
Magnesium carbonate, 29
  and calcium ions, 61
  as component of hard water, 54
  solubility of, 62
Magnesium oxide, 42
Malic acid, 37, 183

chemical structure, 185
Mandelic acid, 184
    chemical structure, 185
Medicines, timed-release, 208–211
Meglumine ion, 129
    chemical structure, 129
Melting point, fusible alloys, 230, 231
Menthofuran, 191
    chemical structure, 192
l-Menthol
    chemical structure, 191
l-Menthol, 191
l-Menthone
    chemical structure, 191
d-Menthone, 191
l-Menthone, 191
Menthyl acetate, 191
    chemical structure, 191
Mercaptans, 168, 180, 181
    chemical structure, 181
Mercury, liquid
    freezing/boiling point of, 112
    use in thermometers, 112
Metal ions
    and food spoilage, 120, 121
    in floor waxes, 125
Metalloanthocyanin, 123
    chemical structure, 124
Metals
    alkali, thermal ionization, 3
    combustible, 39
    fusible, 230, 231
Methacrylic acid (MAA), 224
    chemical structure, 225
Methacryloxypropyl tris(trimethylsiloxysilane)
        (TRIS), 222
    chemical structure, 223
Methane, 60, 63, 180
Methane metabolism
    website reference, 195
Methionine, 60
    chemical structure, 60
4-tert-butyl-4′-methoxydibenzoylmethane,
        163
    chemical structure, 164
Methoxyflurane, 79
    chemical structure, 80
Methyl chloroform, 8

Methyl α-cyanoacrylate, 219
    chemical structure, 220
Methyl methacrylate (MMA), 221
    chemical structure, 222
2-Methyl-2-propanethiol, 168
3-Methylbutane-1-thiol, 169
    chemical structure, 169
Mica, 147, 148
Microencapsulation, 208, 212
Microspheres
    in scratch and pressure sensitive ma-
        terials, 211–214
    silica, in opal, 27
Mineraloids, 26
Mirrors, 143, 144
Miscibility, 16, 212
Molality, 21
Molecular sieves (zeolites), 31, 48
Molybdate, 57
Monochloramine, 182
Monomers
    and copolymers, 125
    and pH, 101
    chain-growth polymerization of, 215
    ester linkages of, 153, 217
        chemical structure, 154, 217
    in aramid polymers, 93
    in cellulose, 209
    in contact lenses, 221–224
    in dicarboxylic acid, 93
    in Kevlar manufacturing, 92
    in polyesters, 96
        chemical structure, 96, 97
    in polyurethane, 125
    in siloxane polymers, 88
        chemical structure, 88
Montmorillonite clay, 31
Morphine
    chemical structure, 177
Myristic acid, 54, 203

Natural gas
    color of flame, 2
    composition of, 180
    impurities of, 47
    odorants in, 168, 180, 181
    website reference, 181
Ninhydrin, 170

chemical structure, 171
Nitrogen trichloride, 182
Nitroglycerin
    chemical structure, 210
    hydrolysis of, to nitric oxide, 210
    oral, 210
Nitrous oxide, 78
    as anesthetic, 81
    chemical structure, 79
Nodihydrocapsaicin, 83
Nonionic surfactants, 97–99
    chemical structure, 99

Octahedral holes, 65, 66
Octyl methoxycinnamate, 162
    chemical structure, 164
2-Octyl-1-dodecanol, 187
    chemical structure, 187
Octyldimethyl PABA, 162
    chemical structure, 163
Octyldimethyl PABA (Padimate O), 162
Oleic acid, 54
    chemical structure, 202
Oleophobicity, 91
Opals, 25–27
    amorphous structure of, 26
    iridescence of, 26
    solid emulsion, nature of, 26
Optical brightener, 158
Optical isomerism, 192
Optically variable ink, 155, 156
Organ pipes, 113, 114
Osmosis, 128, 129
Osmotoxicity, 129
Ovoglobulin, 22
Ovomucin, 22
Oxazole, 159
    chemical structure, 160
Oxidation state, 36, 41, 71
Oxidation-reduction
    chlorine, 71
    hydrogen peroxide, 49, 169
    in discoloration of pigments, 57
    in metabolism, 194
        illustration of, 195
    in metal corrosion, 37
    in solid rocket boosters, 36
    indigo, 46

iron and iron alloys, 43, 44
    magnesium, 39, 40
    ninhydrin, 171
        illustration of, 171
    thiols, 169
Oxidizing agent, 44
Oxybenzone, 162
    chemical structure, 162
Oxygen permeability, of contact lenses,
    222, 223
Ozone, 161

PABA
    chemical structure, 163
Padimate O, see Octyldimethyl PABA
Palmitodiolein, 201
    chemical structure, 202
Panthenol, 187
    chemical structure, 187
Paper manufacture, 149, 150
Partial pressure, 12, 81
Patina, on copper, 36, 37
Pearls, 25–27
Penicillin, 179
    chemical structure, 179
Pentane, 180
Peppermint oil, 190, 191
    composition of, 191
Peroxide compounds, 40, 41
Petroleum gas, 180
PGA, see Poly(glycolic acid)
pH, 19, 72, 73
    albumen, 22
    as factor in timed-release medicines,
        209
    definition of, 72
    effects on iron oxidation, 44
    indicators of, 74, 75
    of drain cleaners, 126
    pool water and chlorination, 70
    soil, factors affecting, 122
Phase separation (coacervation), 212, 213
Phenols, 173
    chemical structure, 173
Phenoxyethanol, 187
    chemical structure, 188
Phosphoric acid, 55
    chemical structure, 55

Photochemical reactions, 71
Photons, 3
Pigments
    in artist's paint, 56, 57
    pearlescent, 147, 148
Piperine, 83
PLA, *see* Poly(lactic acid) (PLA)
PMMA, *see* Poly(methyl methacrylate)
Polarizability, hydrocarbons, 86
Poly(glycolic acid) (PGA), 166
    synthesis, 167
Poly(lactic acid) (PLA), 167
    synthesis of, 168
Poly(methyl methacrylate) (PMMA), 218,
    221
    synthesis, 218
Poly-para-phenylene terephthalamide, 93
Polyacrylic acid, 214
Polyalkylene oxide polyurethane-urea (POPU),
    91
Polycarbonates, 216, 217
    synthesis, 218
    website references, 219
Polydimethylsiloxane, 86
    chemical structure, 86
Polyesters
    ester linkages forming, 153, 217
        chemical structure, 154
    fibers, absorbency of vs. cotton, 96
    in security thread of currency, 152
Polyethylene fibers, 92
Polyhydric alcohol
    cooling sensation of, 9
Polymerization
    condensation, 209
    cyanoacrylates, 220
    free-radical, 214, 216
    ring opening, 167
    solution, 93
Polymers
    biomaterials, 227
        website references, 167
    bullet-resistant fabrics, 92
    contact lenses, 221, 222
        copolymers, 222
        hydrogel, 222
    cross-linking, 125
    heterochain, 217

in biodegradable sutures, 166
in drug encapsulation, 208, 209
in floor waxes, 124
    acrylic polymer, 125
    copolymers, 125
in microencapsulation, 208, 212
in superglues, 219
in thermoplastics, 92
siloxane, 86
water-repellent fabrics, 89
Polyphosphates, as sequestrants, 55
Polysiloxanes, 88
POPU, *see* Polyalkylene oxide
Potassium bitartrate, 14
Precipitates/precipitation, 14, 53
    agents that facilitate, 55
    hard water, 54, 61
    in seashell formation, 51
    lauric acid-calcium, 54
    pH effects, 123
    pigments, 57
Precipitation reaction
    in seashell formation, 51
Pressure-sensitive chemical release, 211
Propane, 180
1,2-Propanediol, *see* Propylene glycol
1,2,3-Propanetriol, *see* Glycerin
Propellants, in solid rocket boosters, 35,
    36
Propylene glycol, 15, 23, 188
    chemical structure, 23, 189
    hygroscopic nature, 24
Pugelone
    chemical structure, 192
Pulegone, 191
Pyrophosphoric acid, 55
    chemical structure, 55

Quartz, 26
Quaternium-15, 98
    chemical structure, 99

Radiopaque media, 1
Raschig process, 182
Red lead, 57
Reducing agent, 41, 44, 49, 194
Reflector material, 65
    website references, 144

Refraction, light, 51
    by solutes in ice, 20
    definition, 150
    in pearlescent pigments, 148
Retention time, 204
Rubrene, 141
    chemical structure, 142
Rust, 43
    as secondary oxidation, 44
    tin plating and, 44

Salcin, 174
Salicylic acid, 172, 184
    chemical structure, 173, 186
Salicylic acid, synthesis from willow bark, 174
Salt, table (sodium chloride), 2, 3
Saponification, 203
Schaefer, Vincent J., 106
Scopolamine, 210
    chemical structure, 210
Seashells, 51, 52
Sequestrants
    in cleaners, 54
    in food products, 120
    in water softeners, 55
Sevoflurane, 79
sevoflurane
    chemical structure, 81
Silica
    amorphous, 26
    as gemstone, 26
    diffraction effects, 26
    hydrated form, 26
    microspheres, 27
    minerals, structure of, 26
Silica gel, 31
Silica glass (vitreous silica), 76
Silicone oil, 16, 86
Silicones, 88
Siloxanes, 86
    hydrophobic nature of, 88
Silver iodide
    dispersing fog, 106
    in cloud seeding, 106
Silver, Spencer, 214
Silvering process, 143
Skunk odor, 168

website references, 170
Smoke, 23, 24
    from solid rocket boosters, 35
Soda-lime silicate glass, 76
Sodium bicarbonate, 63, 64, 67, 116, 169
Sodium borosilicate, 76
Sodium carbonate, 48, 61, 62, 64, 74, 75
Sodium chloride, 2, 3
    crystal structure, 65
    illustration, 66
Sodium hydrosulfite, 46
Sodium hydroxide, 46, 47, 74, 75, 126, 217
Sodium hypochlorite, 70, 182
Sodium ion, 3, 54, 76, 116, 121, 129
    light emittance, 3
Sodium lauryl sulfate, 98
    chemical structure, 98
Sodium silicate, 31, 55
Sodium vapor lamps, 3
Solar flares, 145, 146
Solid emulsions, 26
Solubility
    acid and metal ion in wine, 13
    barium sulfate, 17, 18
    gas vs. solids, 12
    magnesium hydroxide, 19
    of carbon dioxide, 12
    potassium bitartrate, 14
Solubility product, 17, 19
Solutes, depression of solvent freezing point by, 20, 21
Solvents, 10
    in drug manufacture, 176
    in dry cleaning, 91
    organic, 83
    sensitivity of canines to, 178
    volatile, 178
Sorenson, Soren P. L., 72
Specific gravity, 15
Standard reduction potential, 44
Static electricity, and dust buildup, 86
Stearic acid, 54, 203
    chemical structure, 201
Stearyl alcohol, 187
    chemical structure, 188
Stilbene, 159
    chemical structure, 159

Sublimation, 110
Sucrose
    chemical structure, 206
Sulisobenzone, 162
    chemical structure, 162
Sunblocks
    vs. sunscreens, 161, 162, 165
Supercooled clouds, 105, 106
Suprane
    chemical structure, 81
Surface coatings, polymeric, 125
Surface tension, 85
Surfaces
    and adhesion properties, 220
    frictional forces, 86
Surfactants
    amphiphilicity of, 97
    polar headgroups, types of, 97
Sutures, 166, 167

Tartaric acid, 22
    chemical structure, 13
    partial ionization, 13
Tartartic acid
    chemical structure, 185
Temperature
    absolute scale (degrees K), 6
    and 'fire breathing dragons', 49, 50
    and green eggs, 59
    and opal formation, 26, 27
    and tin stability, 113
    and transdermal delivery of medi-
        cations, 210
    anesthetic agents, 79
    baking, 69
    behavior of liquids and gases, 11
    Charles' Law and, 6
    citrus crops, saving, 108
    denaturing of proteins, 101
    density and, 15
    effect on pigments, 56
    expansion due to, 5, 15
    fermentation and, 14
    fire extinguishing and, 63, 64
    flash point, 7
    fog, effect on, 24
    freezing, of water, 20, 105, 107,
        108, 112

ideal, of carbonated beverage, 11
    ignition, of wood, 39
    incandescence and, 110
    melting, fusible alloys, 110
    of Class D fires, 39
    pressure and, 11, 12
    ranges, and thermometer types, 111,
        112
    rate of combustion, 42, 63
    rate of reaction and, 41, 42, 64
    regulation, in seashells, 51
    scales
        Celcius, 112
        Fahrenheit, 112
    solubility of gases, 12
    solubility of solids, 14, 83
    supercooling, 106
    thermal decomposition, 63, 65
    tungsten filament, 110
    vapor pressure and, 7, 9, 10, 67, 68
        graph of, 69
Terephthalic acid, 93
Thermal analysis
    of opal structure, 27
Thermal breakdown
    microcapsules, 212
Thermal coefficients of expansion, 15
Thermal conductivity
    of a gas, 130
    of metals, 144
Thermal decomposition, 63
    reactions involving, 63, 65
    sodium bicarbonate, 63
Thermal expansion
    coefficient of, 112
        liquids vs. solids, 112
Thermometers, 111, 112
Thermoplastic resin, 92
Thermoset plastic (thermoplastic), 216
Thief detection powder, 170, 171
Thiols, 168, 169
    chemical structure, 169
Thymolphthalein, 74
    chemical structure, 74
Tin
    chemical structure, 114
    crystal structure of white tin, 114
    organ pipes, 113

stability of, 113, 114
    standard reduction potential of, 44
Tin Woodman, 43
Tinplate, 43
Titanium dioxide, 8
    crystal forms of, 150
    in paper manufacturing, 150
    in pearlescent pigments, 147
    in sunblocks, 161
    refractive index of, 150
    website reference, 151
α-Tocopherol (Vitamin E), 187
    chemical structure, 188
Transdermal delivery, medications, 210
Transitional metal complex ion, 125
Triazole, 159
    chemical structure, 160
1,1,1-Trichloroethane, 8
Trichloroethylene, 79
    chemical structure, 80
Tridymite, 26
Triethanolamine, 163
    chemical structure, 164, 190
Triethanolammonium, 98
    chemical structure, 98
Triethylene glycol, 23
    chemical structure, 23
    hygroscopic nature, 24
Triglyceride
    saponification of, 202
Trimethylchlorosilane vapor, 90
TRIS, *see* Methacryloxypropyl tris
        (trimethylsiloxysilane) (TRIS)
Tungsten
    ductility of, 110
    filament, 65, 109, 110
    melting point, 110
    website reference, 111
TWEEN 20 polysorbate 20, 99
    chemical structure, 99

Ultraviolet light, 71, 146, 152
    classifications of, 161
    fluorescence of bills under, 153
        website reference, 154
    fluorescence of optical brighteners,
        158, 159
    sunblocks and sunscreens, 161, 162

chromophore absorbtion, 163

Valine, 103
    chemical structure, 103
Vapor pressure
    alcohol, 7
    ethyl alcohol (ethanol), 9
    of drug solvents, 176
    of drugs (anesthetics), 79
    of liquid, 7, 10, 22, 68
    water, 9, 67, 68, 81
        graph of, 69
Vaporization, enthalpy/entropy of, 68, 69
Vinegar, *see also* Acetic acid
    affect on penicillin, 178
    as cleaning agent, 36–38, 52, 53,
        194
    balsamic, sweetness of, 175
    dye protector, 46
    equilibrium constant of, 73
    fermentation to, 176
    types of, 175
N-Vinyl-2-pyrrolidinone (NVP)
    chemical structure, 224
N-Vinyl-2-pyrrolidinone (NVP), 224
Violanthrone, 141
    chemical structure, 142
Viscosity
    of glass, 76
    of lava, 15
Vitamin E, *see* α-Tocopherol
Volatility
    alkali ions in fire polishing, 77
    leavening in baking, 67
    of alcohols, 9
    of anesthetics, 78, 79, 82
    of essential oils, 190, 191
        website reference, 193
    of solvents, 178
        illegal drugs, 178
        paper correction fluid, 8
    of thiols, 169
Volume expansion coefficient, 112

Walker, Edward Craven, 14
Washing soda, 55
    as water softener, 61, 62
Water

bound in opal, 26
content of contact lenses, 221–223
hardness of, 52–55, 61, 62
heat capacity of, 108, 136
polarity of, 95
vapor pressure of, 9, 67, 68, 81
    graph, 69
Wurtzite crystal structure, 106, 107

X-rays
    in CAT scans, 1, 2, 128, 129
    radiopaque media, 17

Zeolites, synthetic, 31
    crystal structure, 31
    website reference, 32
Zinc
    alloys of, 37
    as transitional metal complex ion,
        125
    in batteries, 137
    in floor wax, 125
    in wurtzite, 107
    ion, 125
    website reference, 126
Zinc oxide, in sunblocks, 161, 162
    website reference, 165
Zingerone, 83
Zirconium, 39, 42
    in alloys, 228
    oxides of, 42
Zwitterionic surfactants, 98, 99